FAITH,
MADNESS, AND
SPONTANEOUS
HUMAN
COMBUSTION

FAITH, MADNESS, AND SPONTANEOUS HUMAN COMBUSTION

*What Immunology
Can Teach Us
About Self-Perception*

GERALD N. CALLAHAN, PH.D.

THOMAS DUNNE BOOKS
ST. MARTIN'S PRESS ✄ NEW YORK

THOMAS DUNNE BOOKS.
An imprint of St. Martin's Press.

www.stmartins.com

ISBN 0-312-26807-6

First Edition: January 2002

10 9 8 7 6 5 4 3 2 1

To Gina, my inspiration

CONTENTS

ACKNOWLEDGMENTS

First, I wish to thank my wife, Gina Mohr-Callahan. Without her, I would never have begun this work. It was Gina who gave me the first vision of this book, and it was Gina who held my hand through the last revision of this book. It is because of her that I am a writer. I also wish to thank Dr. Charles DeWitt. Charlie gave me my first opportunity in a laboratory and paid for all the glassware I broke. Whatever skill I have as a scientist, it is largely because of Charlie. And I thank all of those who read this manuscript and cared enough to remind me when I was lost inside my own words—John Calderazzo, SueEllen Campbell, Pattie Cowell, Sue Doe, Mark Fiege, and David Mogen. I want to thank, as well, friends who provided me with literary and technical feedback, and encouragement—Dr. Les Walker, Mary Ellen Corbett, and Dr. Liz Gremillion.

A slightly different version of *Chimera* originally appeared in *Creative Nonfiction*, volume 11, 1998.

Sunday Afternoon on the Island of La Grande Jatte

A woman with a red parasol and a young girl dressed in immaculate white walk side by side through a copse of trees. Discussing, I imagine, the fine French day. Beyond, men lean into the oars of their scull and force their way slowly up the river Seine. Closer by, a man draws meditatively on his pipe, and the smell of the tobacco carries me back to my childhood and the aromas of my father. Sunday afternoon at the river. Dozens of people, going their own way in their own time. No one, it seems, wishes to be anywhere but right here.

In spite of that, that haze of self-satisfaction, there is in this painting a striking intensity. It's apparent in the sky and the river, the trees and the people sitting or standing or walking beside the river. Intensity and more than a little mystery.

In this pointillist painting by Georges Seurat, *Sunday Afternoon on the Island of La Grande Jatte,* nothing is as it seems. If you look too closely, everything disappears. All that remains is a chaotic collection of colored dots. Step back, and the men and the women and the children and the intensity reappear. The effect depends on individual dots, thousands of them, laid one at a time onto the white expanse of the canvas. The painter placed them all within an arm's length of his eyes. But only from a distance do the dots collect themselves into recognizable beings. How Seurat managed that, I can't even guess. But inside of those dots is the secret of the painting's intensity.

In this world, too, the intensity arises from the dots—individual cats, individual men, individual women, individual dogs and trees and girls and boys—dots daubed by who-knows-what hand onto the blue paper of this planet. Friends, pets, gardens, lovers, sons, daughters, husbands, wives. Like Seurat's men and women, the intensity of our lives—the very meaning of our lives—comes from the colors that surround us. Our lives gain clarity only against the backdrop of all those other dots.

It is easy to forget that, though. Since birth, we have been surrounded by others. And because they have been there from the start, we take it for granted that we always have been and always will be surrounded by others. That is the nature of things, it seems.

In fact, that is *not* the nature of things. The nature of things is to mingle, to coalesce, to merge. The only reason that people don't do that as quickly as raindrops on windowpanes is because we have immune systems. In a moment, all of the others in our lives would cease to exist if it weren't for human immunity.

Every day a million, perhaps a billion, living things compete with us for the space we occupy, and these creatures would, if they could, convert the edible parts of us into more bacteria, viruses, parasites, and funguses. The only thing that allows people to survive such an onslaught and to continue as individual men and women is the immune system. Each moment of each day, whether we notice or not, the white cells in our blood and the antibodies in our plasma take on all of the microorganisms we inhale or swallow or allow to pass through the cracks in our skin and dispose of them—efficiently, finally.

If the cells and proteins inside of us didn't do this, humans would quickly become bacteria and funguses, parasites and viruses. Because of immune systems, and only because of immune systems, most of us wake up every morning nearly the persons we were yesterday and the day before.

"He," "she," "it," "I," "me," "you," "them," come to us from our
thymuses, bone marrow, lymph nodes, spleens, appendices, and
tonsils—the elements of immunity. Without these cells and organs, there would be no George W. Bush, no Sinéad O'Connor,
no dogs, no cats, no birds, no slugs. No one. Just a single muddy
pool where the word "us" would suffice to describe all of life.

And the fight for human individuality begins early. Almost
from the moment of conception each of us is at risk. Because of
that, human immune systems develop quickly. Six weeks after
sperm and egg meet, the thymic rudiment—the bit of tissue that
will one day stand between us and all the rest of the world—has
formed. At sixteen weeks, the process of self-discrimination is
fully under way. By then, the thymus sits just above the baby's
heart and is working furiously to form a lasting image of self
inside the developing child.

The young thymus transforms primitive bone marrow cells into
T lymphocytes, a type of white blood cells. Soon it will be the
task of these new T cells to recognize and respond to all that is
not self and to save us from the bits and pieces of the world that
threaten our individuality. A hundred billion dots writhing inside
a human thymus. And from that frenzy comes a first portrait of
self. Nothing even a little like this occurs outside of human thymuses and bone marrow. A miracle nearly. But an inherently
imperfect process.

As the image paints itself, mistakes are made. Some T cells
fail to achieve any functional state and become empty space in
the portrait. Other T cells develop a taste for self and would, if
allowed, destroy us—black spots that threaten to overwhelm the
other colors in the picture. If we are to survive the biological
assault that this world is about to hurl at us, those empty spaces
and those black smudges must be erased. That is the job of the
human thymus. While we are still developing inside of our mothers, our thymuses fill in the empty spaces and erase the black

spots. When the process is done, within each of us there is an image of self, a vibrant portrait made from a billion billion individual lymphocytes. Close up, it looks like nothing more than a chaos of white blood cells. Stand back and suddenly you see a person, unique among us all.

After all of that, and only after all of that, a boy or a girl—a brand-new individual—is born. And so long as his or her portrait of self holds together, one more of us walks the convex paths of this planet.

But throughout our lives, defense remains paramount. Without it, we are nothing more than a fistful of food for bacteria and viruses. Because of that, defense sculpts everything we do. We choose our mates by smell to enhance the defenses of our children. We fear those who differ from us, because beneath the rivers of our minds rumbles the primitive thunder of self-defense. The first messages we get from our mothers speak to us of the dangers of this world—immunological warnings about bacteria and viruses, parasites and funguses, warnings, too, about our fathers. And beyond that, the tones and textures, perhaps even the termini, of our lives are also monitored and mediated by defensive systems. Inescapable machinery, often beyond our senses, grinding out a three-billion-year-old plan written to protect us from all the rest. Written, as well, to protect us from ourselves.

Why we love, why we hate. What we first know of our fathers, how we live and why we die. The light and the dark inside of human eyes, the warmth of human combustion and the chill of rejection, sanity and madness—everything that makes us human men and women is tied to immunity as tightly as to thought. We are humans because we are individuals. We are individuals because we have immune systems.

This book tells that remarkable story—the story of human immunity. What immunologists have found beneath human skin changes much of what we thought we knew about being human.

Hidden inside modern biomedical science, there is a tale that each of us should know. A tale of the starlight and the darkness inside, a tale of the sins of the father and the flame of spontaneous human combustion, a tale of madness and love, of faith and despair. Wrapped up inside of that tale is a portrait of each man and woman in all our wonder, a portrait full of the intensity of life. But like Seurat's painting, if you stand too close, as perhaps we scientists have done, you see only the words, only the dots. You must take a step or two backward if you wish to see the mystery.

SELF-CREATION

I celebrate myself, and sing myself
And what I assume you shall assume
For every atom belonging to me as good belongs to you
I loafe and invite my soul

—WALT WHITMAN

Chimera

Last Thursday, one of those gray fall days when the starlings gather up and string between the elms around here, my first wife, my children's mother—dead ten years—walked into a pastry shop where I was buttering a croissant. She ignored me, which she always does, ordered a plain bagel and an almond latte, picked up her food, and without a glance at me, walked out. The starlings chittered, the day frowned, and I went back to buttering my croissant.

Just after her suicide, I saw this woman often—in towns where she never lived, walking her Airedales in the park, eating poached eggs at Joe's Cafe, sweeping grass clippings from her walk on Myrtle Street, stepping off the Sixteenth Street bus. I caught her smell while walking beside redbrick arches where she and I never walked. I heard her laughing beneath the birches in the park.

We get together less often now. But when we do, like this morning, her image is as vivid as it ever was—her dark eyes as bright, her odd smile just as annoying.

I'm not crazy.

I know it isn't her, this woman I see and hear and smell. After all, she's dead, and I myself gave her ashes to my son. So it is another, a stranger, transformed by some old film still flickering through the projector inside my head. Just someone with a crooked nose like hers, or her perfume. Someone walking with her knees turned in a little too far. Someone else. I know that. But still, every time I see her, it takes all that I have to stay in my chair, or my car, hold on to myself and not run after her calling out her name.

Some of this I understand. When something or someone is suddenly stripped from us, it seems only natural that our minds would try to compensate. Minds do that. If they didn't, we might ourselves be sucked into the vortex. That part I grasp. I'd have thought, though, that in a year or two, the films in my mind would fade and break, and the tear in my life would close and scar like any other wound. And I expected, as the fissure closed, that my first wife would disappear.

I was wrong. Ten years after her death, she's still inside of me somewhere.

———

All the pieces of human bodies fit (more or less) into eleven systems—endocrine, musculoskeletal, cardiovascular, hematologic, pulmonary, urinary, reproductive, gastrointestinal, integumentary, nervous, and immune. So there's a limited number of places where someone could hide something inside a human being. And so far as we know, only two of the body's systems— immune and nervous—store memories, fourth birthdays, or former wives. That narrows it even further.

Most of us don't associate immune systems with hopes and fears or emotions and recollections. Most of us don't imagine anything other than lymph—the pale liquid gathered from the blood—is stored inside of thymuses, spleens, and lymph nodes. The business of immune systems is, after all, not hope, but immunity—protection against things like measles, mumps, whooping cough, typhus, cholera, plague, African green monkey virus; you name it.

But immune systems *do* remember things, intricate things that the rest of the body has forgotten. And the memories stored inside our immune systems can come back—like my first wife— at unexpected moments, with sometimes startling consequences.

My grandmother had a penchant for saving things. She grew up in a very poor family and believed nothing should be wasted. On the plywood shelves of her closets, mason jars that once held apple butter or pickled tomatoes were filled with buttons, snaps, paper clips, and strips of cloth; seashells, rubber bands, pebbles, bobby pins, and cheap shiny buckles—everything she'd ever come across that she thought might be useful someday.

Immune systems do that, too, believe that most everything they come across will be useful again someday. My grandmother used mason jars, immune systems use lymph nodes. My grand-mother gathered beads and hat pins, string and sewing needles— the stuff that fell through the cracks. Immune systems collect bacteria, parasites, and funguses, proteins, fats, sugars, and viruses—the stuff that falls through the cracks in our skin.

Human skin is like nothing else in this universe. It tastes of the sea salt and the iron inside of men and women. Its feel arouses us. Skin is cream, sand, teak, smoke, and stone. Skin is poetry, music, literature, the dreams of boys and men, girls and women. Skin is all of that, all at once. But mostly, skin is what keeps us apart from everything else on this planet, especially everything that might infect, infest, pollute, putrefy, and possess us. First and foremost, it is our skin that allows us to be here as individual men and women in a hungry world. Skin keeps things out—things that would eat us for lunch. And skin keeps things in—things we couldn't live without.

But skin breaks down, gets punctured by knives and needles or scraped off by tree limbs and tarmac, gets cracked, blistered, burnt, and broken. When that happens, if it weren't for our im-mune systems, we'd die—abruptly. Immune systems deal with the things that crawl through the holes in our skin. Immune systems label the intruders as dangerous, round them up, and destroy them. And immune systems remember the things they've seen beneath our skin, because immune systems believe that one day those things will be back.

That's why we get to be adults—immunological memory. It's also why vaccines work. Until a few years ago, children in this country were regularly injected with cowpox, also known as vaccinia virus. Vaccinia virus is very similar to the virus that causes smallpox, with one important exception. Vaccinia virus doesn't cause the disfigurement, illness, and often death smallpox virus causes in humans. But as Edward Jenner discovered in the 1700s, people (in Jenner's case, milkmaids) who had been infected with cowpox didn't get smallpox. A miracle. Immunity to cowpox protects a child from smallpox. That's because even though their personalities are very different, smallpox virus and vaccinia virus have a lot of physical features in common. Immune systems that have learned to recognize and destroy cowpox virus also recognize and destroy the look-alike smallpox virus before it can do harm.

Immune systems remember most of those miracles, too, often for a lifetime. A child vaccinated against smallpox virus makes a more rapid and specific response on a second encounter with that virus than does an unvaccinated child. And the rapidity and specificity of that second response is what saves the vaccinated child's life.

A simple memory of a tiny virus, but a memory powerful enough to have ended the devastating disease of smallpox on this planet. Immunological memory, but in essence no different from the memory that pulls our fingers from the flame a little faster the second time, or the memory that guides the cleaver beyond the scars on our knuckles, or the memory of a first love lost.

The way immune systems remember is extraordinary. Just beneath our skin, there are specialized cells called Langerhans cells. Langerhans cells grab up bits of whatever drops through the holes in our skin—bacteria, viruses, funguses, and so on. And Langerhans cells do this very well. After we are wounded, Langerhans cells begin to gather up the things that crawled in through the tears in our fabric. Then the Langerhans cells make their way to the nearest lymph node. Lymph nodes are little peanut-sized filtering stations strung throughout human bodies. There are literally thousands of lymph nodes in the average man or woman—lymph nodes under our arms, in our necks, beneath our jaws, on top of our intestines—just about everywhere. So there is always one near any hole we manage to punch in our skin. When the Langerhans cells reach those nearby lymph nodes, they hand off bits of what they've collected to another group of cells called T-helper cells. T-helper cells are a type of white blood cells called lymphocytes, and T-helper cells control all the operations of the immune system. Once they see the junk carried to the lymph nodes by the Langerhans cells, the T-helper cells tell another group of white blood cells called B cells to begin making antibodies. Antibodies are proteins that bind specifically to things like viruses and bacteria and help to destroy these invaders. But some of the antibodies made during an immune response do something other than destroy invaders. These antibodies stick to the surface of another specialized cell in the lymph nodes. These cells, called follicular dendritic cells, are to lymph nodes what green glass mason jars were to my grandmother.

The antibodies stuck to the surfaces of follicular dendritic cells also begin to gather pieces of the invader or invaders and hold them there in the lymph nodes. Those bits of viruses and bacteria and funguses do all kinds of things. Most importantly, they don't let us forget what has just threatened us. Follicular dendritic cells and the bits of antigens bound to them help us remember by maintaining a low-grade, barely perceptible immune response against the pieces of the invader inside of our lymph nodes. Then, the next time the same virus or bacterium shows up beneath our skin, that smoldering immune response quickly flares into a systemic immune response—one that is much faster and much more specific than our response the first time we were threatened by this virus or bacterium. Because of that, a second exposure to many infectious diseases is much less likely to be life-threatening.

That—the newfound specificity and speed of the second and subsequent immune responses to the same threat—is *immuno-logical memory,* and it's as powerful as any memory in the brain.

This whole process—from skin to lymph node to follicular dendritic cell—happens each time we are infected. So, after each infection, a few of the bacteria or viruses that infected us are saved in the lymph node where they first arrived, saved on the follicular dendritic cells. By the time we're adults, our lymph nodes—like my grandmother's mason jars—are filled with bits of things we've been infected by: viruses, bacteria, parasites, proteins, lipids, carbohydrates. Just how long lymph nodes hang on to these things isn't altogether clear. But it is clear that lymph nodes keep things for quite a while, maybe even for a lifetime. Lymph nodes are the repositories of our infectious histories: over there's a piece of that flu I had last winter, and beyond that there's a bit of tetanus left over from when I was bitten by a mouse, and here in the foreground is some of the polio vaccine I got when I was six, and behind that lymphocyte—there just below the high endothelial venule—there's another flu virus from an infection I don't even recall, and so on. Just like an attic before the garage sale. But unlike my grandmother, every day, our immune systems sort this growing mass of memorabilia and remind themselves of what they've seen before and what they are likely to see again—memories, infectious memories. Things we mustn't forget.

Mustn't forget, but mustn't hold too close to the surface, either. Just like some of the memories lurking in our brains, an ill-timed immunological recollection can hurt or blind us, sometimes even kill us. These memories we must suppress.

Some of the viruses and bacteria stored on follicular dendritic cells appear to be intact and alive. As long as they are there, the only thing keeping us from having the same diseases all over again is the constant vigilance of our immune systems. Through that vigilance, all of those things hanging around inside us are kept in check, suppressed to the point where they can help us remember, but cannot cause, disease. Memory with a mission, selective recollection and suppression.

But lots of things can distract immune systems: drugs, malnutrition, stress, hormones, age, infection. When these things happen, immune systems can forget about all those deadly things packed away inside of us. Then, like minds in panic, immune systems may get confused, forget which memory to recall and which to suppress, and the past can flare inside of us.

When that happens, our very survival depends on our ability to regain our balance, enhance some recollections, and suppress others. No one knows how humans do this.

And it isn't just our immune systems that store infectious memories. Nervous systems hold dangerous biological recollections as well. A particularly pernicious example of this is shingles, a severe chicken pox-like rash that usually appears across the ribs beneath the arms but also may grow in eyes and lungs. Shingles occurs most often in the elderly.

People don't get shingles unless as boys and girls they were infected with chicken pox. Shingles and chicken pox are caused by the same virus—varicella zoster virus. When we get chicken pox, our nervous systems, and perhaps our immune systems, collect live varicella zoster viruses. Later, when age or illness or depression distracts our immune systems or stresses our nervous systems, or both, some of these viruses begin to multiply again.

Then varicella zoster may blind us, even kill us. This is shingles—a blazing memory of a childhood disease, chicken pox, a thing we wish we could forget.

And there are many others: like cytomegalovirus that causes pneumonia, Epstein-Barr virus that causes mononucleosis, nasopharyngeal carcinomas and lymphomas; adenoviruses that cause acute respiratory diseases like pneumonia, leukemia viruses, some parasites, tuberculosis, HIV (human immunodeficiency virus), and likely many more that we haven't yet discovered. All of them smoldering inside of us, inside of lymphocytes and macrophages, inside of lymph nodes, inside of neurons, waiting for a moment when our immunological backs are turned.

So immune systems, as well as nervous systems, are filled with memories—vivid, painful, sometimes fatal. The fragments of a life lived, bits and pieces of the past. And sometimes, immune systems lose control of this smoldering wreckage and old flames flare anew. Is there a woman, then, living in this ruin, a woman inside of me who walks and speaks exactly like my first wife?

It is, of course, impossible to answer that question. No one understands nearly enough about wives or immune systems. But it isn't an entirely stupid question. Among the things we regularly trade with our wives (and the rest of our families, for that matter) are viruses—colds, flus, and cold sores, to mention only a few.

Enveloped viruses—like those that cause flu, cold sores, and AIDS—are so named because they carry with them an "envelope" of lipids and proteins taken from the host cell (the cell they grew up inside of). Each time we give our flu to our wives or our cold sores to our husbands, we also give them a little bit of ourselves. And when our husbands or our wives get hold of those bits of us, they don't just discard them, they pack some of them away inside their lymph nodes. Pack them away, sometimes for the rest of their lives.

Other viruses move DNA between people. Retroviruses—viruses like HIV—store their genetic information in RNA instead of DNA. When these viruses infect human cells they first make DNA from their own RNA. Most retroviruses then insert these DNA copies of their RNA genomes into the chromosomes of the cells they infect. Sometimes, when new retroviruses are made, these viruses may take with them not only viral genes but pieces of human chromosomes as well. When an infected person later passes the virus to someone else, occasionally some of the first person's DNA gets incorporated into the second person's chromosomes. That means that each time one of us is infected with a virus like this, we also acquire some of the DNA, some genetic piece of the person who infected us. Currently, virologists know of only a few retroviruses, and all of these viruses, like HIV, cause significant human disease.

But the sequence of the human genome reveals many more retroviral sequences than can be accounted for by the few retroviruses we know of. One percent or more of all the DNA inside of humans came from retroviruses. So infections involving retroviruses may be much more common than we currently imagine. Maybe, then, we exchange bits of genes with one another a lot more often than we realize. And there is now evidence that some of these exchanges swap DNA from within the major histocompatibility complex—the segment of the human genome that is most directly involved in immunological individuality, most directly involved in the biological definition of self. Infection becomes communication, memorization, chimerization. At once something fanciful, something mythical, something oddly immunological that leaves each of us holding tightly to pieces of others inside our genetic heart of hearts.

Maybe that seems trivial—a bit of envelope here, a little DNA there. But over the course of an intimate relationship, we collect a lot of pieces of someone else. And a little of each of those pieces is stored on the follicular dendritic cells inside our lymph nodes, and a little of that DNA gets into our chromosomes.

Until. Until the person we've been communicating with is gone, and we stop gathering bits of someone we love. For a few days or weeks, everything seems pretty much like it was. Then one day, for no apparent reason, our defenses slip just a little, and a ghost walks through the door and orders an almond latte.

————

Apart from a few viruses inside of neurons, nervous systems don't appear to store memories in the same ways immune systems do. Most neurologists and neurochemists believe that memory within the nervous system involves something called long-term potentiation, or LTP—a means by which certain nerve pathways become potentiated or preferred.

Every nerve signal that comes to the brain has hundreds, perhaps thousands, of different pathways that it may choose to follow. The path the nerve impulse chooses determines the outcome of the signal—whether we flee or fight, for example. At the head of each of those possible paths there is a gate, called a synapse. The key to the gate is carried by molecules called neurotransmitters—things like norepinephrine and glutamine, dopamine, and acetylcholine. The first time that "two plus two" enters a brain, it may stimulate no response at all; or it may stimulate the release of norepinephrine and cause a recollection of the sound a toy train makes; or it may stimulate acetylcholine release and cause a vision of a ballerina. But after being consistently rewarded for only the word "four" and a vision of four oranges, the gate—the synapse that leads to four oranges—becomes potentiated. Now each time this person hears "two plus two," she responds, "equals four." That's long-term potentiation. In return for it, we get to move on to second grade.

Some of this potentiation we learn, like two-plus-two; and some of it we are born with, like eating when we are hungry. Beyond that, a lot remains to be learned about how nervous systems store and recall memories. A whole lot.

Some neurologists have divided human memory into two broad categories; declarative memory (explicit, consciously accessible memory, e.g., What was the name of the cereal I had for breakfast?) and implicit or emotional memory (often subconscious and inaccessible, e.g., Why was I so frightened by that harmless snake I saw today?). Both of these forms of memory appear to depend on long-term potentiation. But there is evidence for a third kind of memory as well, something I'll call phantom memory. Memories that come from someplace beyond or below declarative and emotional circuits. Whether phantom memories depend on long-term potentiation is unclear.

I'm pretty confident that declarative memory had nothing to do with my first wife walking in on me as I was buttering my croissant last Thursday. I'm less certain about emotional memory. I am deeply intrigued by phantom memory.

People who have had arms or legs removed often experience phantom limbs—a sensation that the arm or leg is still there, sometimes a very painful sensation. This feeling is so real that people with phantom hands may try to pick up a coffee cup just as you or I would. People with phantom legs may try to stand—before their declarative minds remind them they have no legs. Always, the missing limbs seem completely real to these people and as much a part of themselves as any surviving appendage—even when the phantom limb is a foot felt to be dangling somewhere below the knee with no leg, real or phantom, between the ankle and a midthigh stump.

Some of those who have studied phantom-limb sensations argue that these are only recollections of sensations "remembered" from the days before amputation. But children *born* without limbs—children who've never experienced the sensations of a normal foot or hand or leg—experience phantom limbs. Clearly, these phantoms are not simple recollections of better days. Instead, the presence of phantom limbs in children with no real-limb experience suggests that some sort of prenatal image—some template of what a human should look like—is formed inside our fetal minds before our arms and legs develop, before even our nervous systems are fully formed. And this behavior in these children further suggests that if long-term potentiation is involved in phantom memory, then these particular nerve synapses become potentiated well before birth, well before development of the human form is even completed inside the womb. If, at birth, our bodies don't fit this template, don't match our potentiated circuits and the imagery those circuits create in our minds, our brains try to remake reality, twist it until it looks the way our minds say reality ought to look.

It isn't clear where phantom memories reside. Because phantom limbs are often exceedingly painful, physicians have tried to locate the source of the phantom sensations and eliminate them. Spinal cords have been severed, nerve fibers cut, portions of the brain removed. Sometimes, some of these procedures caused the pain to disappear, but it usually returned within a few months or years. And none of these treatments routinely caused phantom limbs to disappear.

Occasionally, some phantom limbs do disappear on their own, though almost never permanently. Nearly always the phantoms will return—in a month or a year or a decade. And when they do, they are just as real as the day they first appeared, or disappeared.

Phantom memories aren't always memories of limbs, either. People who've lost their sight describe phantom visions—not recollections of something they once saw, but detailed images of places they have never been—buildings, burials, forests, flowers.

Similarly, some people who've lost their hearing are haunted by complex symphonies blaring in their ears—symphonies that no one else has ever heard. Beethoven was among those who suffered from such phantom sounds.

It is impossible to say how much of our reality comes to us from the physical world that surrounds us and how much "reality" we create inside our own minds. If, using a machine like a PET (positron-emission tomography) scanner, we were to analyze all of the nervous activity occurring at any given moment inside a human body, no more than a fraction of a percent of this activity would be directly due to input from the senses. That is, only a tiny portion of what our nervous systems are occupied with—and by inference only a tiny portion of our thoughts—results directly from what we see, hear, taste, smell, or touch. The rest of it, the remainder of our mental imaging, begins and ends inside of us.

How that affects our reality isn't clear. But it is clear that much of what originates within us is powerful—powerful enough to fill our mental hospitals with people who see and hear things that the rest of us would say aren't there. Among the sights and sounds that originate within us are our images of ourselves and our realities. Such images are powerful icons, nearly immutable. These are the images of our dreams, our poetry, our theaters, our psychoses. Some of these we are born with. Others take us years to acquire and assemble. In the end, intermixed with the rubble of the archetypical images we began with, there is a picture of us and the world around us—one that suits us and probably only us.

But, if physical reality—the outside world—changes abruptly, it may not be within our power to so abruptly change such deeply rooted images of ourselves and our worlds. A loved one, after years together so nearly a part of us, suddenly and unexpectedly dies. A career of thirty or more years is abruptly terminated by downsizing. Childhood memories that no one can speak of. When that happens, reality itself becomes implausible. Often, our minds can't handle that. At that point, there are only two ways out—madness or a reconstruction of reality. What determines the difference between these two options may be nothing more than the degree to which our reconstruction of reality fits with other humans' visions of reality. At any rate, the only way to staple our imagined realities and "real" realities back together is with a phantom, a bit of virtual reality that reconciles our world and the real world, a stitch to close the void. Our loved one reappears at unexpected and sometimes inappropriate moments, a phantom we have projected onto others to make our reality fit our need. Our stories about who we are change suddenly to fill holes that have opened in our pictures of ourselves—ephemeral patches to mend inner reality. Our past becomes someone else's past, our reality a fiction. Gaps between the hard facts of the real world and the truths we hold inside threaten us. Because we must, sometimes we fill those gaps with ectoplasm.

Are the dead, then, living within my neurons—inside of my own pictures of me?

Images of ourselves—some, apparently, older than we are—are obviously deeply etched into the stones of our minds. These are powerful things that resist change, particularly sudden change. But even the aboriginal portraits of ourselves aren't without seams or cracks. Inside those seams and between those cracks, small forces working over years can introduce change. Time, time in an intimate and powerful relationship, could reshape even our images of ourselves. The changes would be little ones at first—a tiny fissure unmortared here or there, room to include in our self-portraits parts of other men or women, a first vision of ourselves as something more. In the beginning, perhaps, nothing more than the turn of a phrase—one we've admired from the first in the speech of the one we love—now turning up more and more often among our words.

Later, larger pieces of us might be lifted and replaced by whole chunks of another. Husband and wife begin to speak alike, know what the other is thinking, anticipate what the other will say, begin, even, to look alike. Until one day, what remains is truly and thoroughly a mosaic, a chimera—part man, part woman, part someone, part someone else.

Later, if that man or that woman is amputated from us, clipped as quickly and as cleanly as a crushed leg, our minds are suddenly forced into a new reality—a reality without the other, a reality in which an essential piece of us is missing. At that point, our declarative minds would be at odds with our own pictures of ourselves. To rectify that, to reconcile the frames flickering inside with the darkness flaming outside, we conjure a phantom, a phantom to change our world. We force a bit of what is inside outside, out into the real world. We create someone or something that will help us slow the universe for a moment while we repaint our pictures of ourselves—repaint, with a very small brush, a very large canvas.

———

There is a painting by Pierre-Auguste Renoir which I first saw at the National Gallery in Washington, DC. This painting, titled *Girl with a Watering Can,* is filled mostly with the off-whites and intense blues of the impressionist painter. But in the girl's hair, there is a blood-red bow. I've often wondered about that bow and why Renoir put it there. I've imagined the bow was a symbol of the death that begins at each of our births; I've imagined it as an omen of sexual maturity—its pain and its promise; I've even imagined it was nothing more than a schoolgirl's red bow.

But just now, I think the red bow is the other one inside of us, the red one who is probably at first mother—physically, immunologically, and psychologically. The one, too, who is later so many others—grandmother, friend, severed limb, or lost wife.

Renoir placed the bow in the girl's hair, near her brain. I don't imagine, though, that by that placement he intended for us to ignore all the other spots where bits of men and women gather in us.

———

Sitting on the redwood deck behind my house this morning, the air smells of cinnamon and rainwater. For reasons I can't recall, those smells remind me of the Brandenburg Concertos, coffee on Sunday mornings, and the intricate paths of swallows.

Somewhere inside of me, there is a woman. But where she lives and who it was that led her into that pastry shop last Thursday, I've no way of knowing. For one part of me, that ignorance is a gnawing blindness. For another part of me, it is enough to simply know for certain that I will see her again.

Self and Antiself

Even as it begins, the argument is lost. Jennifer, sensing that, smirks at me, and then, like a smith with warm metal, twists my reasoning back upon itself and hammers it closed. I lunge. She parries.

Jennifer is fourteen. We are serious about our arguments. It is our nature and often it is all we share. Neither of us wishes to lose a single point.

Outside the sun is burning just as it has every other day for the last five billion years or so. Opossums are scuffling myopically beneath the dark leafy branches of the avocado trees beyond the window, and the whole wretched scene stinks of the sea.

Inside, my daughter flings another dart. I have lost the thread of the entire argument, but I am angry now, angry with her and her smugness. I reach across the wooden table that separates us and grab my daughter's shoulder. I make some last assertion of my authority. But she smiles, knowing that the moment I grabbed her, I lost. Jennifer smirks once more. Then a worm of reason uncoils inside my brain. I release my daughter's shoulder and I sit back—angry now with myself, but not nearly as angry as I am about to be.

From across the table even, I can see a bruise where I just grabbed Jennifer, a bruise shaped exactly like my left hand. Jennifer notices as well and smiles at me to show that she, too, sees how totally I have failed. My stomach churns.

Jennifer stands to leave, victorious—her bruise her badge. She is wearing a mesh sweater over a spaghetti-strapped cotton shirt. As she turns, I see what I couldn't see before. Where every strand of her mesh sweater was pushed by the chair into her back, there is another thin blue bruise. Her shoulders and her back, a lacework of broken blood vessels.

I am thirty-five years old this morning, sitting at a wooden table off an aging kitchen in southern California. My daughter is standing across the table from me, and for the first time I imagine that she is dying.

In response to her evident fear, I do the only thing I can think of—I lie to her.

"It will be all right," I say, thinking to myself that it will never be. And then I reach to call the hospital.

———

Throughout my life I've relied on the well-told lie to get me through hard times. I've lied about who I was sleeping with to bolster a bad relationship—the lie of protection. I've lied about how I played baseball—the I'm-as-good-as-you-are lie for effect. I've lied about when my check guarantee card expired, so that I could eat—the justifiable lie. I've lied about where I've been—it was just easier. I've even lied at times for the simple pleasure of lying.

And I was good at it. I was a good liar.

I have hated myself for it, though. Hated myself for the ease and the skill with which I lied to others, for the convenience of it, for the dishonesty of it, for the pleasure of it. And at times, I resented those who believed my lies.

I was wrong, though, wrong to hate myself and wrong to resent the gullible. Fabrication is an inescapable part of being human; so is believing our fabrications. Whether we admire ourselves for it or not, whether we believe it or not, we lie—some of us less than others, but we all lie.

Every foundation ever erected to support human beliefs and human behaviors is riddled with rotten concrete—spots where logic is conveniently moldy, gaps that we could never find the glue to fill, but gaps that we came to overlook because we couldn't resist the final construction. Into those gaps, we squeeze the mortar of our lies.

Even science—in fact, especially science—the chalice of human truth, is littered with half-truths and outright lies. Few scientific manuscripts published in the last twenty-five years accurately relate how research is actually done. Instead, most accounts seek to leave the impression that all of the scientific issues were appreciated from the outset, and none of the experiments failed. When in truth, many experiments were poorly designed, sloppily executed, or overlooked. And most of the time the whole idea of what was truth changed dramatically from the beginning to the end of the project. None of that is told in the final story.

And even the finest of scientific theories is full of holes. But acknowledging these holes is incompatible with the process. So, by mutual assent, we ignore them.

Gregor Mendel, the nineteenth-century Austrian monk who laid the groundwork for modern genetics, and his contemporary Louis Pasteur, the father of microbiology and the theory of infectious diseases, both lied about the results of their research. The lies are apparent in the near perfection of their reported results. Peas don't behave as precisely as Gregor claimed they did to prove his theory of inheritance. And Louis's flasks must have been contaminated with bacteria and molds more often than he reported, because the techniques for sterilization available to him were simply not good enough to so routinely prevent so completely contamination of his culture media.

Both of these eminent scientists falsified their data—deliberately misrepresented their experimental results to support their preconceived notions. And even if they themselves didn't doctor the data, as some have claimed, they were both good enough scientists to have recognized the ruse. Neither, though, ever complained about the neat rows of numbers they made up or were given. This is considered the cardinal sin for a scientist, intentional or collaborative misrepresentation of data. But we overlook Pasteur's and Mendel's transgressions, because we believe, first, that the preconceptions of both men were right, and second, because modern genetics and microbiology are founded on those results.

Science—the pinnacle of our pursuit of objective truth—is itself rooted in opinion, personal bias, and sometimes lies. I don't believe that detracts much from science. I believe science is still a powerful means for looking deeply into the things of this world, including ourselves. And most of the fabrications of scientists are self-correcting. But science is a human endeavor, and humans lie. And we use our lies like taxis to take us places where the world is as we wish it to be.

We lie, too, about lying itself because we've been taught to believe that confabulation is wrong. Even one of the great Judeo-Christian commandments, one of our commonly held "truths," admonishes us not to lie. The problem with that is it implies that we know when we are lying; it implies that we know how to stop; and it implies that truths themselves did not begin as lies.

In spite of our aversion to the sport of lying, it is our lies that speak most honestly about who we are. Truth tells us nothing about a man or a woman. We all, more or less, agree about truth—all of us. All of us, that is, except for a few outliers, who can hardly be considered human. And once we've all agreed on it, true or not, "truth" loses all anthropological interest. Everyone agrees. Everyone speaks of truth in more or less the same way. If we wish to discriminate between individuals, truth is useless. What distinguishes us—each from the other—is the lies we tell.

No two of us tell exactly the same lies; no two of us lie for exactly the same reasons; no two of us are bound in the same way by the lies we have told. The essence of an individual human being, the black hole at the center of self, is a stack of fabrications—an intricate interweaving of misperceptions, misinformation, misunderstandings, and outright lies. Our individuality is bound up inside of our lies.

But that isn't because of some aberration or some character flaw. It is the human way. It isn't nature that abhors a vacuum, it's humans. We humans don't believe in the limits of human knowledge, even temporary limits. We don't accept the spaces between what we know and what is. So we lie. We lie to fill in those spaces and smooth the fabric of reality. Otherwise this universe, this life, would be unmanageable, overpowering, and terrifying. We lie to make it manageable—all of us.

Nor were humans the first to resort to untruths as a means to an end. Lying is an old family tradition. Chimpanzees will lie about where food is hidden if the trainer, after being directed to the food by the chimp, routinely takes and keeps the food. Killdeers and plovers lie about broken wings and busted bodies to lure predators from their defenseless young. Butterflies, moths, grasshoppers, beetles, caterpillars, spiders, and fish pretend to be things they are not. One becomes a leaf, another a bug-eyed snake simply to frighten those who might have them for dinner. It's an old and grand tradition.

But we humans are best at it. Though language or self-awareness is often offered as "the" quintessential human quality, we as a species might as easily be distinguished from other species by our facility with the lie. We will and do lie for almost any reason at all to anyone who will listen to us, including our selves. It is our way.

This has several important implications. First, because of its ubiquity, it is tempting to speculate that the human lie is no simple convenience and that the roots of untruth lie deep within each of us. Second, it means that everything that each of us was ever told about this universe and our relationship to it—by others or by ourselves—is suspect. And finally, it's likely that I am lying to you right now.

———

In Jennifer's case, my lie turned on me. In spite of my belief that I was deceiving her, I had told her the truth.

After diagnoses of leukemia, then lupus, a false-positive anti-DNA antibody test, a bone-marrow biopsy, and a visit to a pediatric oncologist, my wife and I are told that Jennifer has neither leukemia nor lupus. Leukemias are devastating malignancies of the bone marrow, often fatal. Lupus, or systemic lupus erythematosus, is an autoimmune disease that attacks many parts of its sufferers, including their own DNA. We are greatly relieved that Jennifer's disease is not lupus.

Next the doctor tells us that Jennifer is suffering, not from leukemia or lupus, but from something called idiopathic thrombocytopenic purpura, ITP. "Idiopathic," loosely translated, means "we have no idea how this happens." "Thrombocytopenic" means low levels of platelets in the blood. "Purpura" means bruising.

"Your daughter has a lot of bruises and almost no platelets in her blood. None of us has any idea why." Over the years, physicians have discovered that such diagnoses leave most parents dissatisfied, some angry. So the disease is now called "autoimmune" thrombocytopenic purpura, AITP. But it remains mostly idiopathic.

"Thrombocytes" is another name for platelets. Platelets are pieces of megakaryocytes, and normally human blood is full of platelets. A healthy adult human has about 275,000 per cubic millimeter of blood. That day, Jennifer had about eighty. One of thrombocytes' main tasks is to plug leaks in capillaries, the small blood vessels that abound in skin. Apparently, we damage those vessels often, but we never notice the damage, because platelets plug the leaks as fast as they appear. When the number of platelets drops to eighty, though, there are not enough to plug the leaks left by an angry father's squeeze, not even enough to control the damage inflicted by a chairback on a child's skin—then the damage each of us has done is apparent.

It isn't the chair's fault. No one could imagine that the chairback intended to bruise Jennifer. No one. In fact, it really isn't anyone's fault, is it? Things like that just happen. Don't they?

Finally, the doctor recommends that we do nothing for Jennifer. Nothing is something I do reasonably well, but something that is not easy for my wife, Gina. She wishes she could do more. Six weeks later Jennifer's platelets rise to 150,000. After ten weeks, they are at 265,000—normal.

———

I am elated by Jennifer's recovery, but I remain dissatisfied with the diagnosis. I am reminded of a Monty Python skit in which a man is trying to obtain a license for his pet bee (perhaps it was actually half a bee). The government clerk in desperation hands the man a license and says, "There. A bee license." The man takes a look at the license and is instantly outraged. "Why, you've simply crossed out 'cat' and written in 'bee.' I want a proper license." Or something like that. Physicians, it seems to me, have simply crossed out "idiopathic" and written in "autoimmune."

Immune systems are, by their very nature, destructive. Destruction is what immune systems do. Autoimmune diseases arise when a human immune system mistakes human flesh for the enemy. Why immune systems do that is still open to question.

Autoimmune. The word itself is a deception. It's not only nearly meaningless, it's too clean—like boraxed porcelain. It doesn't sound like knuckles splintering. It doesn't smell of self-inflicted wounds. It doesn't speak of the simple horror of watching your daughter disassemble herself. And there is no warning in that word about how you will feel when you realize that it was you, after all, who gave your daughter the genes she needed to unbolt herself.

Lupus. Now there's a word.

———

Oddly, young women, at least after the age of sexual maturity, are much more likely to develop autoimmune diseases than are young men. On average, women's immune systems fail to discriminate between what is self and what is not self about five times more often than men's immune systems do. About five times more often, a woman's immune system fails to notice that her knees are not the enemy, that she cannot live without her kidneys, that her skin is her ally. About five times more often, a woman's immune system systematically dismantles a human body. Ten to twenty women develop systemic lupus erythematosus for every man who gets the disease, three to five women for every man with rheumatoid arthritis, three women for every man with scleroderma; same for multiple sclerosis, five times more women get Hashimoto's thyroiditis, and young men rarely get AITP.

No one has ever offered a fully satisfying explanation for this inequity.

———

Among its permanent collection of artworks, the National Gallery of Art lists twenty-two self-portraits. Only one of these was painted by a woman—Judith Leyster, in Amsterdam in about 1630.

Within the painting, Judith is completing a painting of a foppishly dressed young man playing the violin. She is seated, head turned toward the viewer, arm raised and resting on the corner of her chair. Her mouth is partly open, and she looks as though she is saying, "Good Lord, how long has it been? Please, sit down."

The Fine Arts Museum of San Francisco lists over 266 self-portraits in its collection. Among all of those paintings, etchings, lithographs, photographs, woodcuts, etc., there are only eight self-portraits by women. All eight of those were carved or drawn or etched by the same woman—Käthe Kollwitz.

Käthe Schmidt was born in Königsberg, East Prussia, in 1876. She met and married Karl Kollwitz in Berlin, where she was studying art in 1889. Throughout her life she was a staunch socialist and champion of the working class. In 1914, her son, Peter, died fighting for Germany at Diksmuide on the western front. She never fully forgave herself for his death. She had, after all, given her blessing to his enlistment. Afterward, much of her art turned to the horrors of war and its impact on women. "No longer diverted by emotion," she once said, "I work the way a cow grazes."

Käthe Kollwitz's eight portraits are startling. She is twenty years old in the first painting, seventy-one in the last. She smiles in none of them. She looks directly at the artist, herself, in only two. Each is stark and disturbing, but those painted after her son's death are the most disturbing.

In these two large international collections, it is nearly a hundred times more likely that we will find a self-portrait by a man than it is that we will find a self-portrait by a woman.

It may be that the scarcity of self-portraits by women reflects the frequency with which women were allowed to be artists or how seriously women's paintings were taken or preserved. Perhaps. But it might be, too, that women have truly been less frequently drawn to self-portraiture. It might be that some women are less certain about their images of themselves, less taken with the prospect of laying their portraits onto canvas, or wood, or stone.

———

The frequency of autoimmune disease and sparsity of self-portraiture by women could be linked, possibly to one another and almost certainly to women's self-perceptions. After all, autoimmunity is a failure to discriminate between self and not self, and self-portraiture is a concerted effort to isolate the self from the sea of "other" that surrounds each of us. Self-perception as self-destruction—immunologically, artistically. How could something like that happen?

Our world, the world we gather through our eyes, is a world of individuals, the vision of a pointillist—each tree, each crane, each fern a point in the scatter plot of life. Because that is what we see, that is what we believe—the endpoint of eighteen billion years of universal will is a sea of individuals. That is simply how things are.

But of course, that isn't "simply" how things are. Individuals aren't a given. Individuals—you, my daughter, your mother, the neighbor's dog—exist only so long as each has an understanding of what it means to be an individual, only so long as each has a clear image of self. A moment of forgetfulness, and one of us is gone. One moment there is a hippopotamus, the next there is only the vital frenzy.

That essential image of self is the gift of the immune system. To verify that the immune system is at the heart of it, it is only necessary to watch what happens when immune systems fail. HIV causes human immune systems to fail, and the consequence is the erosion of the individual and the birth of a community. The process may take ten or more years, but it culminates in the destruction of a human being and the birth of a colony of living things. A virus empties out a part of the immune system, the self-portrait frays, and abruptly a man or woman is gone.

Three billion years of evolution have polished immune systems into very lustrous and hard points. Immune systems are good at what they do. If they weren't, I wouldn't be writing this, you wouldn't be reading it.

But it isn't the same for each of us. Some immune systems have different assignments. Women's immune systems, for example, must identify and destroy every living thing that enters women's bodies, save one—fetuses. And fetuses, because half of their genes come from their fathers, are guaranteed to be foreigners—different from their mothers in many of the same ways that bacteria and viruses are different from human mothers, different in the same ways that other people's kidneys and livers are different from mothers, different in all the ways that normally elicit immediate immune destruction.

Men need make no such distinction. Everything that infects a man is the enemy. Women must be more selective. This places an odd demand on their immune systems, chinks in the armor of self, smudges inside of self-portraits. Somehow, after a thirteen-to-forty-year battle against every foreign thing that ever entered their bodies, women must tolerate nine months of parasitism. And throughout that nine months, women must resist the powerful immunological urge to destroy that parasite.

During pregnancy, systemic lupus erythematosus flares, rheumatoid arthritis recedes. Shortly after pregnancy is when autoimmune thyroiditis most often develops. During pregnancy, fetal cells in the placenta secrete an enzyme, inidoleamine 2,3,-dioxygenase, or IDO. IDO destroys tryptophan, an essential amino acid. Without tryptophan, the mother's immune system falters. The battle subsides. The parasite prospers. The picture of self blurs. After pregnancy, some women experience postpartum depressions so severe that the identities of both self and child collapse. Shortly after delivery, women with this illness have tried to kill themselves, their children, both. It is a time of great personal and immunological crises.

Of course, every woman doesn't face these crises. Some women never become pregnant. But evolution has no way of knowing who will and who won't become pregnant. For the sake of simplicity, evolution imagines that all women will become pregnant, imagines that all women will house fetuses, believes that every portrait of a woman must include the face of a child.

———

Women in this world face another odd task, the task of self-creation. Once the placenta is fully formed between fetus and mother, the mother begins passing antibodies to her daughter or her son, passing her strength to each of them, passing it equally. Those antibodies protect what is soon to be a newborn child from all who have threatened the mother—bacteria, viruses, funguses—things the mother has survived, things that may threaten the child. Equally, each of her children is told about where the mother has been and what she has seen.

At birth, things change in two ways. First, the mother's antibodies move into her breasts, where they are passed to nursing children for nearly a year. And second, the equality ceases.

From birth, perhaps even sooner, men (or those who will become men) are bathed in images of men. And the role of men in this world is laid out on day one. We are told that our lot is to rule. We are to acquire skills, marry, procreate, amass wealth, and rule. It isn't complicated; every boy understands it. And if by chance any of us missed any part of it, television, movies, novels, plays, paintings, poetry, newspapers, magazines, and sports remind us every day. The message is: men were made to rule.

For many women there is no obvious counterpart—there is no gift of an equally obvious role. What this world whispers to these baby girls is: "You are to marry, you are to bear children, you are to serve." Even as children this message is empty for most women. Men hear of strength and power—gifts for the self. Women hear of beauty, marriage, and service—gifts for others. Doctoring and lawyering—professions of power and self-service—are for men. Teaching and nursing—professions of giving to others—are jobs for women. Women know from inside their own genes that there is more to life than that. But often, no one—not even mother culture—is going to tell a woman (or those about to become women) what more there is. Because of that, many women are left to define themselves relationally. The only gift is the gift of context—a woman's father, her family, a woman's husband, her in-laws. Foreground and background intermix. Self-portraits become family portraits.

———

And there is more. The proximal cause of AITP seems to be antibodies, antibodies against platelets. But the existence of such antibodies inside an otherwise healthy woman tells us something has gone terribly wrong. Somehow this woman's immune system has mistaken her platelets—part of her own blood—for something horrible, something predatory. Because of that mistake, this woman's immune system begins to systematically destroy her platelets, and for that mistake she may die. The same mistake is rarely made inside of men.

Human immune systems have an odd way of creating an image of self. Immune cells, lymphocytes, have no way of knowing what is self and what is not—anything that binds to a lymphocyte's surface is assumed to be dangerous, and it is destroyed. A little like a land mine, *any* footfall—friend or foe—will trigger it.

Because immune cells are like that, our bodies have evolved

means to protect us from ourselves. Inside of the human thymus, immune cells that recognize elements of self are destroyed as they arise. And the few self-reactive cells that survive this ruthless selection and escape the thymus are destroyed elsewhere in the body or forcefully suppressed. Any immune cell looking for a piece of self is destroyed or suppressed, and what survives this jihad are only those cells that recognize and respond to nonself, at least ideally.

The immune system, and in particular the thymus, charcoals in the spaces between, gouges out what is self and leaves behind only the negative image, the nonself image. A little like the sculptor creating a wooden carving of a bear, who says her task is simply to remove from the wooden block everything that is not bear. But more directly, like a sculptor who sets out to create a wooden image of nonbear by chipping from the wooden slab everything that *is* bear. Our immune systems create the opposite of self-portraits—pictures of the world from which every trace of self has been plucked. A nonself portrait, an antiself portrait.

Human minds appear to do the opposite. Human minds scrape out what is not bear and construct the positive image—in living color, a picture of self. *Cogito ergo sum.*

Matter and antimatter. Neither image possible without the other. Just as darkness cannot exist without light or light without darkness, self and antiself are irreversibly intertwined. Each creates the other. The head of a lion, the tail of a serpent. Both essential. Each the other's flawed mirror. Both vital. Both fragile. And clearly both lies—a white one that claims to know everything that is of the essential self; and a dark one that claims to see everything but self—neither, obviously, possible. But it is essential that we believe both are true.

We know our brains lie to us regularly. The blind spots in our vision, created where the optic disc interrupts the retinas at the backs of our eyes, are filled in for us with pictures of what our brains imagine might be there, but often isn't. In some people the left side of the brain (where most of the language centers are) will make up completely fictitious tales about what the right half has done with the left hand. As with optical illusions, our brains at times will insist that things are right before our eyes when there is nothing there at all, or tell us we've a leg or a hand where there is only empty air.

And our immune systems tell us that their portraits of self are true (when they aren't). So, at the very heart of self there live two lies. We are a storytelling species. It is our stories that sustain us. Even the most rudimentary of human functions requires that we imagine we know and care about who we are. Likewise, any infection, no matter how minor, would be enough to destroy us without an iron-clad image of nonself, and the ability to act on it. We need both stories and the lies they conceal.

———

We don't yet know just how either the neurological or immunological self-images are constructed. But we know how easily they may be destroyed, how easily we may be made to see the falsehood of both, and we know how dangerous that is.

Neither do we know for certain what triggers autoimmune diseases like rheumatoid arthritis or multiple sclerosis, but it appears that each is often preceded by an infectious disease, apparently viral, that causes a set of "flulike" symptoms followed by a sometimes sudden, sometimes insidious, assault on self.

One possibility is that the virus looks to the immune system a little like self, and when the immune system strikes at the infectious virus, the blow falls on self as well as nonself. In the confusion of the instant, the immune system's image of self fails, because its image of nonself is muddied with bits of self. The lies we have been telling ourselves for years are no longer enough because the immune system doesn't believe they are lies. So it lashes out at us, and continues to lash at us as though we were the most dangerous of life forms, lash at us sometimes until we die. Because of a lie.

The mind, too, believes in absolute truth, and trusts implicitly in its image of self. Until. Until one day we notice that some bit of reality is completely at odds with our mind's I. Until one day we notice that the stories we have been telling ourselves are just that—stories. In that instant, our minds perceive the lie. Then, like immune systems, minds may fail to distinguish threat from empty gesture, friend from fiend, life from death. And like immune systems, minds may then destroy us.

Lies. But essential lies. Lies without which none of us would last long. But dangerous lies that can fail us, can destroy us. Falsehoods we depend upon. And like light and dark themselves, falsehoods that only intensify one another.

———

It is hard to be a human being. It is hard to live a lie. It is harder, even, to live two lies. But it is hardest of all for women. First, because in between the negative image of the immune system and the positive image of the nervous system, there sits a defenseless child—the fetus, the future. That, of course, changes everything—but not in any way anyone can predict or pretend to understand. A woman's negative picture, her antiself-portrait, must somehow exclude the child's father-to-be. Evolution has foreseen that a portion of that man will come to reside inside of this woman, and evolution will not allow her to destroy him. And a woman's positive image, her self-portrait, must somehow include the face of the child-to-be, perhaps the faces of many children-to-be. These she must protect as though they were her. Evolution has seen to that, too. Finally, for some women, there is a third lie, a lie that further twists the first two, the cultural lie—self-portrait as family portrait.

An unborn child shreds the fabric of nonself. Biological and social thumb strokes blur the perspective in the painting of self. Background and foreground intermix. Images fade, paint cracks, and canvas curls.

———

Jennifer was fourteen, awash with hormones, when her immune system attacked her. And though I don't recall ever telling her that her role in this world was to serve others or that her pictures of herself must always contain the faces of children, I did squeeze her shoulder hard enough one morning to leave an ugly bruise. If I was capable of that, who knows what I might have said to her when I wasn't thinking or I was angry. Who knows what lies I might have told her then. What lies I might be spreading just now.

It appears true, though, that self begins in at least two different places—inside the mind and inside the thymus. But the rest of it? Jennifer was in the midst of a very confusing time—biological and psychological—when she began dismantling her own platelets, when her immune system mistook her platelets for God knows what. Nothing like that ever happened to her again. And Jennifer is now finishing a Ph.D. in philosophy—still arguing, but with normal numbers of platelets dancing in her veins. So it might be that that was how it all happened one morning in southern California. Immune systems do work like that, minds, too, and fetuses, and societies certainly create problems for both minds and immune systems. So it might be that Jennifer's immune system tried to kill her because for one moment—a moment of intense confusion—at least one of her lies failed her. And in the instant of unmitigated truth that followed, her immune system moved to destroy her.

It might be true, but that's the problem with truth. It's dangerous. Truth is bald-faced, brutish, and simple. Sometimes it is even fatal. The art is in the lie—how we weave it and how it sustains us.

Eating Dirt

The words are old, so old that no one remembers who first spoke them. But the story remains. It is Holy Week, Good Friday in fact, and Don Bernardo Abeyta, a Penitente, is walking the low hills above El Potrero in what is now northern New Mexico. As he walks, the cross-shaped scars carved into his back burn with a pain that is nearly unbearable even for Don Bernardo. Still, the young Penitente is performing his deserved penance. No one knows for which sins he sought his God's forgiveness that evening, and no one remembers just how he sought to relieve himself of those sins. But it is easy to imagine him, like others of his order, lashing himself methodically as he walks, praying for forgiveness—his own and all of mankind's. The evening is cold, the air a pudding of piñon and juniper. Overhead, the stars have closed into hard points that light only the way into nothingness. His sins and his church weigh heavily upon Don Bernardo this evening, and that weight rides in his forearm as he brings the salted yucca braid down hard across his right and then his left back and shoulder. His breaths come in sharp rasps in rhythm to the blows of the short whip. Red welts thick as ropes rise where the weed strikes him. Blood seeps from the bruised flesh and trickles slowly down his small brown back. Don Bernardo is serious in his efforts to purify. But something distracts him, a bird's call perhaps, or a dry twig snapping somewhere in the brush ahead. As the bird cries, or the twig snaps, Don Bernardo drops his arm. When his arm falls, his head rises from his chest. There, in front of him, the earth itself, the red dirt at his feet, burns with the flames of a thousand stars. For a moment, he sees nothing, then the light reaches slowly beyond the scorching pain rolling across his bare back and touches his throbbing mind. The young Penitente drops to his knees and begins digging in the rough red soil with his bare hands. Nearby, the Sangre de Cristo Mountains rise like stone teeth blotting out the stars, and the Santa Cruz River reaches quietly for the Gulf of Mexico.

Finally, breathing heavily, his hands and arms red from soil and blood, Don Bernardo unearths a crucifix. As tall as a man, it stands in the middle of the glowing earth. A cross of wood, green and simply carved. And nailed to that cross he finds a Christ of an unearthly beauty.

Don Bernardo rises from his knees, looks—at the twisted junipers and piñon washed by the light of the cross, the blackened back of the Sangre de Cristos—and then he runs to tell his neighbors of his wonderful discovery.

The story says that Don Bernardo and his neighbors returned and knelt before the cross. Selves were blessed and holy words whispered by each of those there. Then one of the men ran to tell the nearest priest, Father Sebastian Alvarez in Santa Cruz.

When Father Alvarez arrived, he lifted the crucifix from the ground and carried it to the church in Santa Cruz. There he placed the crucifix into the alcove above the main altar. Everyone rejoiced. Mass was said, tears fell, and blessings were given. Then the people were sent to their homes, where they told wonderful tales of the *Milagro* of Santa Cruz. But the next morning when Father Alvarez walked into the church, the miraculous crucifix was gone from its place above the altar.

The Reverend Father gathered the townspeople for the search. But to everyone's surprise the cross was found quickly. To everyone's even greater surprise the cross was found in the earth ex-

actly where it had first been found. Twice more the crucifix was
taken from the earth and hung above the altar in Santa Cruz.
Twice more the cross disappeared from the church and reap-
peared in the earth of El Potrero.

In the end, Don Bernardo Abeyta built another church to
house the cross and the Savior—El Señor de Esquipulas. This
church was built over the hole where El Señor was first found.
This church was named El Santuario de Esquipulas.

It is a good story, one that bears retelling. But it is still a story,
and all of us know how stories sometimes wander off on their
own. So no one can say for certain how much of what the story
carries on its wooden back actually happened. It was Holy Week,
though, and near the vernal equinox—a time when almost anything can happen. Nearly two hundred years later, there is a
church that stands in what was once El Potrero. In that church,
there is a crucifix of a most unearthly beauty. The church is
known now as El Santuario de Chimayo. And people from all
over the world come to visit this small adobe mission-style
church. But for most of these pilgrims, the cross and El Señor
are not what they have come for. Most of those who have come
have come to eat the dirt beneath the church.

———

The church in Chimayo, El Santuario, is shaped more or less
like a cross, as are most Roman Catholic churches. This formal
architecture is most obvious in the basilicas of Rome with their
well-defined narthex, nave, bema, apse, and transept. In all of
these, the altar sits nearly at the intersection of the arms of the
cross. In the Santuario de Chimayo the cruciform architecture
is a little distorted. But the narthex, nave, and transept are clear.
The back of the transept is filled by the most unusual reredos I
have ever seen—floor-to-ceiling panels painted in the same reds
and ochres, the same sepias and rusts that fill the nearby hills.

The arm of Christ crossing the arm of Saint Francis, the Holy Cross, a shaft of wheat, a dark cluster of grapes all vividly portrayed. In the center is the apse and inside of the apse is El Señor, much as He was two hundred years ago, I assume.

Which in and of itself makes the trip to Chimayo worthwhile. But most people who come here pay this image of the Savior only perfunctory respect as they pass beyond the communion rail and through a low portal to His left. There the stucco and adobes open onto a small anteroom that is filled with candles and crutches, metal braces and rosaries with beads of stone and acorn, santos and small *retablos* of simple work, stout canes and plastic statuary, silk flowers and baby shoes covered with the dust of the centuries. Gifts. Gifts left here by those who brought them but no longer need them. Gifts from the cured and shoes for the Santo Niño, the child Christ who, according to local legend, wears his own to rags while walking the nearby hills. But again, interest here is only passing. What most have come for is through the portal at the opposite end of this room There, through another low stone arch, is a place where the floor of the church has been removed (or, more likely, was never filled), the *pósito*. Through that opening in the floor, anyone who wishes can see the miracle of El Santuario de Chimayo. In the center of that room, beneath the mud flooring, there is, to no one's surprise, dirt—the deep red dirt of Chimayo. This is what the pilgrims have come for. This is what they eat.

———

Eating dirt is practiced on every continent in this world and has been for centuries. But, so far as I know, there is only one other Catholic shrine in all the world where the faithful come to eat dirt. That shrine is in Esquipulas, Guatemala. Esquipulas is near the Guatemala-Honduras border. The great basilica there houses the Black Christ—an image of Christ like no other, or nearly no other. This Christ was carved from dark woods like balsam and orangewood to more closely resemble the natives of this region of Central America. And since then, centuries of candle smoke have even further blackened the wood and the cross, until they are both the color of fertile soil. Because of this ebony Christ, every year on January 15, there is a great pilgrimage to Esquipulas. People come from as far away as Mexico and Costa Rica to worship the Black Christ and to pray for healing. Many of them take away with them small embossed cakes made from the fine kaolin clay found in the nearby hills. These cakes, this dirt, the pilgrims eat.

Inside of the shrine at Esquipulas, the Black Christ hangs from a blackened green cross embroidered with gold leaves. The cross in the alcove above the altar in Chimayo is green as well, and is embellished with gold leaves. The Christ on the cross in Chimayo has for as long as anyone remembers been called El Señor de Esquipulas. No one has ever been able to explain that—explain how the imagery of a green cross and the practice of eating dirt could have been transferred across thousands of miles and between two cultures, or how those who brought the cross and the tradition could have left everything in between untouched and arrived with both intact in Chimayo.

———

In the legends of almost every society, human disease comes from the gods. Jaljogin in the Punjab of India hides in the streams and wells and preys upon women and children. In Sumerian mythology, it was Namtar who carried pestilence from the underworld. And in the Bible it was Jahweh who brought down disease, frogs, and the death of their firstborn sons onto the idolaters and slavers of Egypt—and before that, visited the diseases of the world upon Adam and Eve for having dared to eat from the Tree of the Knowledge of Good and Evil.

Our legends tell us, if we care to listen, that disease is divine. Sometimes punishment for things we did, as in the Garden, sometimes discipline for things we should have done but didn't, as in Egypt, but it is always a gift. A gift from the gods.

———

When I started college, I wanted to be chemist. In particular, I wanted to be a protein chemist. I was enthralled with the forces of entropy and how they drove proteins irreversibly toward the matters of the living. Enthralled.

But every time I tried to establish my career in protein chemistry I found myself more securely in the grip of pathology. I enrolled in a graduate school of biology, but no one would pay me to work in biology (mucking about with thermophilic blue-green algae). So after two quarters, I left school and took a job digging ditches. A highly educational and sometimes satisfying job. But that wasn't to be my future either. The dead and the diseased were still reaching out to me. One day, as I dug in the rocky soil of northern Utah, my wife got a call from a man I had sought employment with months before. That man was a pathologist. Shortly thereafter, I was working among the bits and pieces of others' lives. Two years into that program, the pathology department at the University of Utah developed its own Ph.D. program, and I was irretrievably afloat on the sea of the sick.

Throughout graduate school, I tried hard to study protein chemistry. I took courses that no one else in the pathology department took. I hung out in the biochemistry department. I rigged elaborate experiments and broke expensive equipment trying to probe the chemistry of proteins. But in between partial differential equations and physical biochemistry, I kept walking into rooms where someone's mother or daughter, someone's father or son was spread across a steel table like an empty hand. Other days I wandered into out-of-the-way closets where glass jars full of human hearts and livers, kidneys, lungs, brains, tumors, toes, eyes, and ankles lay labeled with black ink and racked like sweet pickles and okra from last year's garden. Yellowed by formalin or phenol, these fragments of folks, more or less like me, fascinated me, drew me in as I had never been drawn by the living. Hours I'd spend staring into those jars into other eyes and hearts and wondering about their lives and mine. Somewhere along the line, I was infected.

When I left graduate school, I tried to find a job in a laboratory exploring protein chemistry. And I found that sort of job—kind of—in a department of pathology at a private research foundation in a small white building made even smaller by the vast blue of the Pacific Ocean at the very edge of America. I spent ten years there.

When I left La Jolla, it was to take a job in a department of pathology. Once again, happily ensconced between the dead and the dying, my career sputtered, grants went unfunded, and papers went unpublished. The administration complained some about that, but the dead seemed okay with it.

———

There is no deeper way to gain respect for the operation of a healthy human body than to walk among the disabled and the moribund. The things that diseases do to human beings are beyond belief. Infectious diseases are among the ugliest of them all—leprosy, tuberculosis, smallpox, plague, elephantiasis, trypanosomiasis, gangrene, necrotizing fasciitis, leishmaniasis. They cripple, disfigure, blind, blacken, bloat, blister, bloody, and ulcerate the living. Infectious diseases are among the most lethal of diseases as well. Worldwide, infectious disease is still the leading cause of premature human deaths. Because of that, immunologists speak of medicine in terms of war—the war on cancer, the battle to recover, defense, killer cells, smart bombs, destruction, antibodies, effectors, enemies, and allies. Of course, there is no war, but it clearly reassures us, though I'm not sure why, to speak of this as though it were a military operation.

And like all wars, in the war against disease, what we would most like to do is vanquish our enemies, or, short of that, isolate ourselves so securely that our enemies could never reach us. To this end we have bactericidal countertops, bactericidal soaps to wash our dishes and our bodies, we have bleaches and crib mattresses that kill bacteria, we have toilet-bowl cleansers that destroy most all living things, we have paints and toothpastes that kill germs, we have diapers that are bacteriostatic and deodorants that kill everything that grows on skin. And we tell our children to wash their hands after every trip to the bathroom and never, *ever* eat dirt. Dirt is filled with the things the gods created to destroy us. Dirt is the enemy, and this is, after all, war.

It's us against them. It's our immune systems (our biological armies and weapons) against the bacteria, the viruses, the prions, the parasites, and the funguses that would, if they could, eat our brains out and dance inside our emptied skulls. That's war. But a most unexpected thing happens when the "enemy" in this particular war is eliminated from the conflict.

———

Rabbits' immune systems develop in an unusual manner. Unlike humans, rabbits begin life with a relatively undiversified immune system. That means that the antibodies and T cells—the "weapons" of the immune system—can only recognize and destroy a very few of rabbits' enemies. Also unlike humans, most of the diversification of the rabbit immune system occurs after birth in the gut-associated lymphoid tissues—places like the appendix.

Because all of the immunological diversification that occurs in rabbits occurs after birth and in known places, rabbits have been especially useful for the study of how the immune system develops in mammals.

Interestingly the single greatest aid to immune development in rabbits is infection. Rabbits that are protected against infection from birth—taken from their mothers by cesarean section, raised inside sterile containers, breathing sterile air, eating sterile food—never develop functional immune systems, never acquire the ability to recognize genuine threats to their lives and eliminate those threats. If these rabbits are exposed as adults to normally innocuous microorganisms, they quickly succumb to systemic infections. What they lost as pups, they cannot reacquire as adults.

The "enemy" somehow generates a rabbit's army. In the absence of infection no troops are ever mustered for the defense. Rabbits that escape the sting of infection—totally avoid the soiled and the septic—become permanently defenseless. Infection is not the enemy. Each of us is, in fact, infected by about 10^{14} bacteria even as we are sitting here reading this. That's a lot, a whole lot, of infection that seems, for the most part, to do us no harm. The word "enemy" hardly seems appropriate here.

————

The Environmental Protection Agency estimates that children in this country consume, on average, somewhere between two hundred and eight hundred milligrams of dirt per day. That's about one-eighth to one-half teaspoon, on average. Undoubtedly, some days our children eat a lot more than that, and some days I'm sure they eat none at all. Still, that doesn't seem like a lot of dirt. Nevertheless, we parents have tried for years to put a stop to it. We just don't like it. But in spite of our aversion to dirt, I don't know of an instance in which anybody ever tried to raise a normal human being under absolutely sterile conditions. Though it seems parents would prefer that, such experiments tend to make people testy. But mice have been raised under sterile conditions; so have guinea pigs and rats. And in each case the animals' immune systems fail to develop normally. Lymph nodes—the little repositories of lymphocytes and the site of most immune responses—don't achieve the right shape or composition in these animals, aren't filled with the appropriate B cells and T cells, and cannot initiate normal immune responses. Reexposure to infection later in life doesn't work. There is a window when infection drives the immune system toward the proper end. After that, mice and rats, rabbits and guinea pigs are at the mercy of this infectious world.

As I said, the same experiments haven't been done with children, but it is likely the results of such experiments would be no different from those obtained with other animals. Children from large families, especially children with lots of older brothers and sisters, are much less likely to develop asthma, hay fever, and eczema. West African children who have had measles are half as likely to develop allergies as children who never had measles. Italian students who have recovered from infections with hepatitis A have fewer and less severe allergies than fellow students who were never infected by this virus. Children with Type I diabetes (an autoimmune disease) are significantly less likely to have had infections before their fifth birthdays than healthy children of the same age. The list goes on.

The immune systems of children who have escaped certain types of infections are apparently altered, permanently. These children's immune systems are much more likely to either miss an infectious agent altogether, overreact to nonthreatening substances—like pollen or peanuts—or completely mistake the children themselves for the enemy and initiate autoimmune attacks on their own tissues. Infections that occur early in humans' lives also appear to aid in the development of normal immune systems.

So it seems, for humans, like rabbits, there is a window in childhood when our experiences, our infections, change everything, once and for all. Inside of that window, infection causes lymph nodes to enlarge and restructure themselves, to organize into cortices and medullae, into primary lymphoid follicles, and develop T and B lymphocyte-rich regions of immune competence destined to someday be germinal centers where our defenses will muster and the real battles be fought. It is a moment in our lives when the simplest and lowest forms of life—the infectious, the parasitic, the septic—alter who we are. During these brief encounters with other life forms, we are changed—changed in both body and mind, changed in the shape and the tone of our immune systems, changed in the ways we defend ourselves, and changed in the ways we perceive ourselves. Even as it arises, self is reshaped by what we have touched and what has touched us in return.

When that window closes, nothing can put the immune system back together again.

Several experiments suggest infections with mycobacteria may be most important. Mycobacteria are a large group of small bacteria, most of which cause no apparent disease. A few strains, though, cause horrible diseases such as tuberculosis and leprosy.

Mice injected with ovalbumin (the major protein in egg white) often develop allergic responses to ovalbumin. But if the mice are first infected with mycobacteria and then injected with ovalbumin, no allergies develop. Children infected with *Mycobacterium vaccae* are more likely to develop protective immunity when vaccinated against tuberculosis than are uninfected children or children infected with other strains of mycobacteria.

People in countries where parasitic infections of the gut are common are much less likely to develop inflammatory bowel diseases. And people intentionally infected with parasitic worms often recover from the same inflammatory bowel diseases.

And so on. To some of us immunologists it appears that early infection of children with certain strains of mycobacteria aids in the development of normal immune systems, normal sense of self, and normal defense of that self. Mycobacteria are found in large numbers in soil. Dirt. And it seems that those of us kept from this dirt lose our ability to recognize certain dangerous organisms as a threat, lose our ability at times to clearly discriminate between self and not self, and lose our ability to distinguish the fatal from the innocuous. Those of us raised in the dirt muster powers of defense unimaginable to the uninfected.

———

Though every male zebra finch has learned to sing by the age of eighty days or so, female zebra finches never learn to sing. Not a note. It isn't because they don't wish to sing; it's because they can't sing. They can't sing because, as they grow, their own hormones destroy the parts of their brains needed for song making. You can make female zebra finches sing in a laboratory if you fill young female birds with male hormones. But outside of a laboratory, no one has ever heard the song of a female zebra finch.

Male zebra finches learn to sing by listening to their fathers' songs—which, by the way, are all nearly identical. But young male finches can only learn those songs during a certain period of their lives—from about twenty to forty days after hatching. If male finches are raised where there are no adult males to teach them, the young birds never sing a true song. Instead, these birds develop "isolate" songs, garbled, nearly unrecognizable vocalizations unrelated to the songs of their species. And once a male finch has learned a particular song, he can never learn another.

All of this—this learning, this changing, this singing—happens because the size of finch brains and the neural circuitry in those brains are changed by listening to songs and by the act of singing itself. When the young birds are sung to by their fathers, the high vocal centers and the robust nuclei of the archistriata—regions of the finches' brains—enlarge and reorganize; enlarge by as much as tenfold, and simultaneously rewire themselves, changing their circuits in irreversible ways to accommodate song. In birds that are never sung to, no such enlargement or rewiring ever occurs.

A simple song, a lyric, an air, one finch's notion of music changes the shape and the weave of another's brain. How remarkable.

———

In the world of animals, such learned vocalization is rare. Most animals, whether they have ever heard them or not, are able to make the sounds of their species—dogs bark, coyotes wail, bears growl, eagles scream, cats meow, cattle moo, elk bugle, and swans whistle—even if they've never heard another of their species make a single sound. Only songbirds, some whales and dolphins, a few bats, and apparently humans are known to *learn* the sounds of their species.

As they learn, though, both a miracle and a tragedy rise inside the mouths of those who learn to speak. Shortly after birth, most human infants can recognize and discriminate among the phonetic units of all human languages. *All* human languages. It seems that is our gift. But even by the age of six months, and most clearly by the age of twelve months, while infants continue to respond to and discriminate among the phonetic units of their own language, what these children have heard has changed them. By twelve months of age, most infants no longer even recognize phonetic units that are not used in their own language. No longer even *recognize* the multitude of sounds that are never spoken in their native tongue. No longer even hear the sounds of others as they roll from foreign lips, simply because the one-year-old has never heard them before.

The words we have listened to since birth have changed our brains, changed them so that certain sounds now just disappear somewhere inside our skulls, simply disappear. Adult English speakers cannot discriminate Hindi consonant/vowel combinations. American English speakers cannot discriminate between the Spanish *b* and *p*. Adult Japanese cannot distinguish between American English *r* and *l*. An entire symphony of human speech has simply vanished. And we have no way to even estimate the magnitude of that loss.

Before that, even before a year of age, there is much we have already lost. We learn to hear best the rhythm and tempo of our mothers' language, and we learn to hear that before we are born. *In utero* exposure to the sounds of our mothers' speech patterns changes who we are. Because of this, by the time we are born, each of us has already developed a preference for our mother's spoken language, already prefers our mother's voice over those of other women, and immediately prefers those children's stories our mothers might have read aloud during the last three months of their pregnancies. Before birth, we know her voice, we have heard the things that she has said, and because of that music, we are changed. That is the miracle.

But because of that miracle, before we are even born, the music has robbed us of some of our hearing. That is the tragedy. Our mother's own sweet voice has narrowed our ability to perceive the very world we are born into. The sound of our mother's words, the stories she read aloud to us, have not only changed the way we think but changed the way we *can* think, changed even what we will ever be allowed to hear. Changed forever.

———

The instant we are born they are there to greet us. And if that greeting is delayed a few weeks or months we will never be the same. Without the lyrical and the infectious bits of this world, it seems we are lost. Of course, the definitive experiment would be to raise a child completely isolated from song and infection. Even though that experiment has never been and most likely will never be done, there may be other ways to assess the consequences of a childhood filled with absolute loneliness. Humans have been raised without ever (or nearly ever) hearing a human word—raised in near-absolute isolation from the storm of words that engulfs most of us at the instant of birth. Such protection has surprising consequences.

The two most studied of these cases involve one child who grew up in the forests of late 1700s France, and a second discovered in Temple City, California, in 1970.

Near the end of the eighteenth century, in the region of Aveyron, France, a young boy was found wandering in the woods. No one was ever able to ascertain how he had come to be in the forest, but he had been there for a long time. He was somewhere near thirteen years old, but only four and one-half feet tall and covered with scars. Those who worked with him called him Victor. Everyone else called him "The Wild Boy of Aveyron."

Victor was a bit small for his apparent age, filthy when they found him, and spoke not a word. A young physician, Jean-Marc-Gaspard Itard, took an interest in Victor, and for the next five years Itard devoted himself to the study and training of the child. Under Itard's tutelage, Victor learned social skills, how to dress and bathe, what to eat. But in spite of all the doctor's efforts, Victor never learned to speak more than a few words, and those he often used inappropriately. There was nothing wrong with Victor's hearing or his larynx. He simply seemed incapable of learning the rudiments of human speech. While that remained a great frustration to Itard, his experience with Victor eventually led the doctor to establish a school for the deaf. And from that work ultimately came Gallaudet, Seguin, Montessori, and schools throughout Europe and the U.S. for the deaf and the retarded.

But Victor never spoke, or at least never spoke as he might have if he had heard words earlier in his life. For Victor, the window had closed. In 1828, at approximately forty-three years of age, Victor died. Dr. Itard died ten years later. In spite of the intimacy and intensity of their years together, their relationship ended without either ever having really spoken to the other.

———

In November of 1970, a nearly blind woman, her aging mother, and a young girl accidentally walked into the General Social Services Office in Temple City, California. The mother had been looking for services for the blind, but her poor eyesight and fate led her to social services instead. The social worker who helped them was immediately drawn to the child. The girl was small and stooped. She walked with a halting gait and a tortured posture. She drooled and spat frequently. The social worker excused herself for a moment, and while the three waited, she told her supervisor that she thought someone had come to them with an unreported case of autism and she thought perhaps they should look into it. The supervisor followed her back to the desk where the three people waited. When she saw the girl, the supervisor wasn't sure about autism, but she was sure something was wrong and agreed they should follow up. That follow-up gradually uncovered one of the most bizarre instances of prolonged child abuse ever reported.

The girl, though apparently near age thirteen when she was examined, weighed only fifty-nine pounds and was only fifty-four inches tall. She could control neither her bowel nor her bladder. She acted as though she had never seen solid food, she couldn't chew, and she had considerable difficulty swallowing anything other than liquids. Her eyes wouldn't focus on anything more than about fifteen feet away. She had a callus that extended clear around her buttocks. She was insensitive to heat or cold, and she spoke only two words—"stopit" and "nomore." Her name, her mother told them, was Genie.

Genie's mother was blind; her father was insane. For thirteen years, Genie had lived with nearly no human contact. Much of that time, she sat naked, strapped to a child's potty chair. The rest of the time, she lay straitjacketed in a cage made from her crib. Her father spoke to her only in growls and barks. Her mother was kept from her, and when Genie made noises of protest, her father beat her. Following her initial examination, the police took Genie into custody. A short while later, Genie's father killed himself, saying: "The world will never understand."

Genie was raised in near complete isolation—no socialization, no vision beyond the narrow confines of her room, few sounds, and almost no words. Social scientists who learned of the story were infatuated with the possibilities.

So at age thirteen years and seven months, Genie first tried to learn a human language. For years, scientists and teachers worked with Genie, but she never gained the sense of language that any of us who were taught language as children acquire without even thinking about it. She was able to use her limited vocabulary to describe some of the things that had happened to her, including things that happened to her before she knew words. Ultimately, though, she was never able to overcome the horrors of that life. Because of that, she eventually landed in a home for the severely retarded and later simply disappeared.

It appears that the Penitentes of northern New Mexico and southern Colorado evolved from the Third Order of Saint Francis. The First Order of Saint Francis was founded in 1210 by Saint Francis of Assisi. The Second Order of Saint Francis, the Poor Clares, was founded by Saint Francis' disciple, Clare of Assisi, in 1212. The Third Order of Saint Francis was established in 1218 to allow laypersons to adopt the vows of the Franciscans without the additional vows of silence and the bond of Holy Orders. The habits of the Third Order are little different from those of the Penitentes.

The Penitentes of northern New Mexico and southern Colorado gained permanence and a realm of their own when Spain withdrew the priests of the First Order of Saint Francis as the Spanish empire's holdings diminished following the Mexican Revolution. Out of necessity, the Penitentes then assumed the roles the priests had vacated—comforting the sick and dying, performing marriages, celebrating masses, overseeing burials and baptisms.

Los Hermanos Penitentes, Los Hermanos de Luz, or simply the Penitentes was and is a society little known to those outside its membership. It is known, though, that initiation into this society involves wounding the initiate's back in the shape of a cross and repeated scourging. The man who administers the wounding and the scourging is called the *sangrador*, and it is believed by all involved that the more energetic and ardent the *sangrador* is in his job, the greater will be his reward in the next life. It is known, too, that all or nearly all of the penitent's later wounding and scourging is self-inflicted, using short whips made from yucca strips softened by soaking in salt water. And it is known that the religious and social devotion of these men is tightly interwoven with Spanish and Catholic traditions and equaled by few others of any order or commitment. Their wounds, they believe, strengthen them and help to atone for their sins, make them worthy of another life—a life of adoration with the Savior of this world.

Don Bernardo Abeyta, the man who found El Señor de Esquipulas buried in the hills of El Potrero, was a Penitente. Since the practices of the Penitentes reach their zenith during Lent and especially during Holy Week, Don Bernardo was likely scourging himself most zealously the Good Friday night he found El Señor in the luminous red dirt beneath the solitary junipers. It was Don Bernardo of the Penitentes, as well, so the story says, who built a chapel over that dirt and raised El Señor into the apse of that chapel for all to see.

In a small room just off that chapel now hang the crutches, the braces, the broken candles, the canes, the crucifixes, and the rosaries of the once ill. Nowhere, perhaps, is it more apparent than here—the church of Don Bernardo Abeyta the Penitente, the self-flagellant—that as long as something is a whole thing it can be only one thing. But once it is broken, it can be anything at all. Any thing at all.

———

There is another old story that is spoken of among the people of Chimayo. Some believe that it was a statue of the Christ Child, Santo Niño, that came from the hole in the ground that night as Don Bernardo Abeyta walked the hills of El Potrero. Not a crucifix, but a child. The Christ Child, Santo Niño de Atocha. In fact, that is what most people today will tell you of the hole beneath the church in Chimayo. Who knows? But this belief in the Santo Niño once led to an odd religious war that changed Chimayo. The story tells of Severiano Medina, a man terribly jealous of Don Bernardo and the financial successes of El Santuario de Esquipulas. Supposedly, Medina went all the way to Mexico, to Zacatecas, to find a statue of the Santo Niño de Atocha. When Medina returned to Chimayo, he built his own chapel, next door to El Santuario de Esquipulas. Medina's chapel drew many of the faithful. But he had been fooled by the men who had sold him the statue in Zacatecas. The Santo Niño Medina had returned with was actually a German papier-mâché doll forced into a sitting position to look like the Santo Niño de Atocha.

That discovery, however, seemed to bother no one, and Medina's chapel thrived and his Santo Niño threatened to steal the faithful from El Santuario. To save El Santuario de Esquipulas from oblivion, its owners, the Chavez family, found another Santo Niño and announced that now the true Santo Niño de Atocha resided along with El Señor in El Santuario de Esquipulas. The new Santo Niño was placed on a low wall in the room with the *pósito*—the hole that leads into the dirt beneath the church. El Santuario was saved, even though the new Santo Niño de Atocha was also a fake. This Santo Niño is actually a copy of the Infant of Prague, clearly recognizable from the globe and cross in the Santo Niño's right hand.

But the faithful still come to both chapels. And most of those who come to El Santuario de Esquipulas these days believe it is the Santo Niño de Atocha, not El Señor de Esquipulas, that is responsible for the miracles that are said to have happened in this little church. In the midst of these assorted claims to the miraculous, with His globe and cross in hand, the Santo Niño simply stands in that small room in Chimayo and stares at the dirt beneath the church.

———

If their fathers don't sing to them, young songbirds never sing. If we parents never speak to them, our children never learn to speak. If sons and daughters aren't infected by the vermin that surround us, children's immune systems never learn the infectious nature of this world, never learn the intricate pathways of self and self-defense. And if a man had not been methodically wounding himself one night in the hills of northern New Mexico, there might be no room full of unneeded cast-off crutches and canes and braces and slings inside a small chapel in Chimayo, New Mexico.

"Whatever doesn't kill me, makes me stronger." Nietzsche said that. He was nearly right. Unquestionably the hurdles and the holes this life pitches at us, the things that crawl inside of us, the sounds we hear, the smells we take, and the fell slap of the whip have made us richer and stronger. Look at the legacy of Don Bernardo's yucca braid and at the ultimate consequences of Dr. Itard's futile struggles with Victor. Look at the finches. Look at our children. But Nietzsche was wrong on one count, dead wrong. Isolation doesn't often kill human beings, but it doesn't strengthen them, either; it cripples them, especially children—cripples them irreparably.

We are born with two selves, but what we become is the measure of our experiences. We are our experiences. The things we perceive make us who we are. Our perceptions change the size and shape of our brains, they change the architecture of our immune systems, and they change the circuitry of our nervous systems. What hasn't killed us has made us stronger. But it is true, too, that the things we fail to perceive—perhaps inescapably fail to perceive—ensure that there are some things we can never become. If too little is given to us, if we are saved from too much of the filth and pain and din of this world, our paths may so narrow that we are strangled by them. No matter the mettle of the selves we began with, the paucity of our experience, the limits of our sensations can cripple us, can so wither us that what remains is barely human.

The dirt, the breadth and depth of the dirt in our lives, the hard whip of this sweet blue world, and the music of human sensation are our salvation. How we grow from infancy into adulthood, how firmly and finally we acquire our selves, what doors open and what doors close are consequences of where we have been, what we have smelled and eaten, what we have seen and heard and touched—all of it self-inflicted, all of it essential, and all of it, no matter how hard we may try, too little.

Self-Defense

The man next to me spits with a certainty that I envy. The sun moves from behind a cloud. And as we watch, his mucus rolls itself into a pale ball in the yellow dirt. That seems to please him. He smiles and pushes at it with a toe. Then while the wind works at his corn-silk hair, he drags a sweatered arm across his busted mouth and spits once more. A poorer effort, and the last of it channels down his chin. He leaves it there, leans back against the old bench, and glares at the people walking past us.

The wind shifts, and I can smell him now—ambergris from a whale's gut, goat urine, a year or more of walking overcoated down sun-bleached streets, nights of fear beyond imagining, human semen.

"Goddamn," he mutters and layers over all the rest a lungful of corpse breath. "Goddamn."

We might be here. I mean, *truly* here on this bench, the old man and I. Or we might not be. Perhaps one of us is dreaming the other.

"Goddamn."

Certainly none of those who pass by our bench testifies to our existence. Each man or woman looks carefully away, at the pavement or the sky, as though ignorance stinks a great deal less than we do, and a handful of coins and keys is handful enough.

"Goddamn."

"Goddamn it all to hell," and where his lips collapse between his broken bottom teeth he hisses and spits once more.

"This wasn't my idea, it was yours from the start," he says to no one.

"You better not let me see you 'round here or I'll break your damn face." And then in spite of the eighty-five-degree temperature, he pulls his overcoat tighter. "Bitch," he mutters.

And he begins to cry. As he sobs he works at the buttons on his sweater with his black and gray hands.

"I didn't mean it," he says to the sidewalk. "Please don't call me that, you know how I hate it. You know how mad it makes me." Now he is crying in earnest, pulling at his hair and his nose and his ears, punching at his eyes.

I reach to touch his shoulder. To help, I think. Just before I touch him, he suddenly turns my way, as though noticing me there for the first time. Immediately, his gaze hardens, he pulls back. A cup of horror fills behind his eyes. Suddenly, he jerks himself from the bench, raises his arm before him as though I might strike him, and moves off, backstepping. When he reaches the paved walk, he turns. Once, twice, three times as he retreats, he glances back, glaring over his shoulder, and then he disappears among the passersby.

———

Ten years ago, maybe more, I read an article in which a man proposed that the immune system—the organs and the cells that defend us against infectious diseases—was an aspect of the brain. He proposed, in fact, that the immune system was a "mobile" aspect of the brain evolved to detect those things we couldn't hear, see, touch, taste, or feel—things like viruses and funguses. It seems, especially for an immunologist, heretical to suggest that the immune system is actually part of something else. But what led this otherwise cautious scientist to make such a startling proposal was the unusual amount of interaction that occurs between the brain and the immune system. First, all the organs of the immune system—like the thymus, where lymphocytes learn the secrets they must know to destroy the enemy, and the lymph nodes and spleen, where immune responses take place—are connected by nerves to the brain. Second, many, perhaps even all, neurotransmitters produced by neurons (the heart and soul of the nervous system) act on both neurons and lymphocytes (the heart and soul of the immune system). So the nerves that attach to the thymus and the lymph nodes to the brain do affect the processes that occur inside these organs. On top of that, many of the cytokines (molecules made by one cell to communicate with another) produced by lymphocytes act on neurons as surely as they act on lymphocytes. So the thymus and lymph nodes, even individual lymphocytes, may be sending messages via the peripheral nerves to the brain as often as they are sending messages to other lymphocytes. Many of the hormones produced by the adrenal glands—in direct response to signals received from the brain—change the behavior of lymphocytes. For example, when the brain determines that something that has been seen, felt, touched, tasted, or heard is cause for anxiety, the hypothalamus—a part of the brain—releases a compound called corticotropin-releasing hormone. That hormone causes the pituitary—also a part of the brain—to release adrenal corticotropic hormone. This hormone is secreted directly into the blood and ultimately causes the adrenal glands to produce cortisol. Cortisol does lots of things. One of those things is to interact with specific receptors on lymphocytes and to suppress immune responses.

So, because of something we've seen, or heard, or touched, or tasted, or smelled—a word spoken against us, perhaps—our brains may take control of our immune systems and shut them down.

This network is the hypothalamic-pituitary-adrenal axis. The HPA. Though maybe we should call it the hypothalamic-pituitary-adrenal-immune axis. The HPAI. I don't know where "axis" came from, but it remains. The rest of the name came from the fact that many sensory signals, input from eyes, ears, etc., deliver signals to the hypothalamus at the bottom of the brain. Often, these signals arise in the amygdala—two almond-shaped slivers beneath the cortex of the brain and connected to many areas in the brain, especially sensory-processing areas. The amygdala determines whether things sensed pose some sort of threat—whether something we saw should make us mad or glad, whether something we felt should make us angry or relieved—and sends signals to (among other places) the hypothalamus. The hypothalamus then signals the pituitary, either directly or with hormones, and the pituitary releases hormones of its own that cause the adrenal glands to release another set of hormones that change all sorts of things, including blood pressure and respiration rate and heartbeat, appetite, libido, general awareness, and pupil diameter. But most especially the hormones produced by the adrenal glands change the immune system. And they change the way this system reacts to the things that threaten us.

Cells of the central nervous system (the brain and spinal cord) also produce cytokines—soluble proteins like interleukin 12, tumor necrosis factor, and alpha-interferon—that act directly on lymphocytes and change the course of immune and inflammatory responses.

Noises we hear can change the rate at which our lymphocytes divide, change the frequency with which our lymphocytes kill. The stress of exams can change the character of the antibodies we produce, and physical restraint can change the way our white blood cells ingest bacteria, the way our T lymphocytes respond, and the way other cells, called natural killer cells, defend us

against infected and tumor cells. *Our thoughts change our immune systems.*

Sometimes it's the other way around. When the immune system believes that we are threatened by bacteria or viruses or something else, the lymphocytes that respond to that threat secrete their own cytokines, cytokines like interleukin 6 and interleukin 1, both of which act directly on cells of the brain to do things like stimulate sleep, suppress our sexual drives, alter our moods, raise our temperatures, change our blood pressures. *Our immune systems change our thoughts.*

The slippery things that make us happy when we are sad, angry when a moment ago we were forgiving, or wake us when we are sleeping—things like endorphins, enkephalins, growth hormone, alpha-melanocyte-stimulating hormone, neurotransmitters, hormones, glucocorticoids, epinephrine and norepinephrine—also change how our lymphocytes respond to bacteria or funguses. And cytokines like interleukin 1 and 3 and 6—things that our immune systems make when they are angry—change our minds.

There is an intimate and continuous dialogue between the cells of the nervous system—including those of the brain—and the cells of the immune system. This dialogue is as immediate and intricate as any that occurs between two inmates in a single cell, as intimate as any that occurs between the two halves of one brain, as intimate as any that occurs among the cells of a single organ. In spite of all of this, I think the man who wrote that the immune system was part of the nervous system was wrong.

The immune system isn't part of the brain. The brain is part of the immune system. Mind is an arm raised against things too large to be destroyed by antibodies and cytotoxic T lymphocytes, the microscopic weapons of the immune system. The immune system is for plague, tularemia, toxoplasmosis, measles, mumps, and chicken pox. Mind is for bears, coral snakes, sharks, snapping turtles, wife beaters, and Buicks.

The immune system was first, by a billion years or more. Viruses and bacteria and funguses threatened all of our ancestors long before sharks and Buicks. Long before. Minds, at least human minds, are just by-products of a much older event—an afterthought—added a billion years later when our microscopic actions were no longer enough to protect us from a macroscopic world.

Why else would one human being hate another simply because of his skin color or her love of women? Why else was the man who was just next to me so certain that given one chance I would hurt him? Why, other than fear, defense, immunity?

The immune system—brain, neurons, thymus, lymph nodes, lymphocytes, defense. A system finely honed to defend us, as necessary, from all the other parts of this world. Part of that system is human mind. Our minds were built for defense.

That's what I'm wrestling with this morning as the man running from me, from *me,* takes a final fearful glance over his shoulder, then disappears into the crowd.

If our brains are only concerned with protecting us, what does that say about human behavior? Is there anything more than defense? I mean, isn't that what moved the man who just ran from me? Isn't that what I'm doing, trying to justify my behavior? Isn't that what every one of those who passed us by as we sat on this bench and spit and cried was doing? Isn't that why they pretended not to notice us? Defense? Self-defense?

———

It is one month earlier. My mother is sitting on the padded seat of her bentwood rocker. Her head slumps a little forward. Her eyes are nearly closed. Beyond her, I can see the enclosed courtyard and a white-haired woman pushing an aluminum walker along the paved path. It is a sunny day. There are finches in the fruit trees.

"Did you hear about my parents?" she asks me, lifting her head.

"Hear what, Mom?"

"They died, you know, both of them. Killed in that El Niño thing." She purses her lips and shakes her head to show her disappointment.

My mother's father died in 1949, her mother in 1962, or thereabouts. Neither death, as far as anyone has been able to ascertain, had anything to do with unusually high subsurface ocean temperatures in the tropical Pacific.

"I know, Mom."

"They're both dead, aren't they?"

"Yes, Mom."

"Do I have any brothers or sisters left?"

"No."

"I didn't think so." Mom's oldest sister, Ethel, died two years ago. Ethel was the last of my mother's ten sisters and brothers.

"I'm the last, then." She shakes her head again, then looks at the picture on the bedside table. It is a picture of her and my father taken just before their fifitieth wedding anniversary, nine years ago—two years after my mother's cerebral aneurysms. The portrait has been retouched to make it appear that her left eye follows her right.

"You know I love my husband Wayne, but I still resent how he always insisted that I accompany him on all of his fieldwork. I didn't care much for that." My father was the superintendent of maintenance at a petroleum refinery. He did nothing "in the field."

Mother stares out the window, taken for a second with the dance of light among the clouds.

"Did you know I worked at a soda fountain once?" Mother tilts her head and adds, "Shorty's. Shorty made a pass at me once. I bit him."

"Yes, I remember you telling me."

"You know I had to carry the money to the deposit every night. I had to walk down the alley. . . ." And she's off. Again. I've heard this story a thousand times, ten thousand times—this story, and maybe ten or fifteen others, that she tells and retells and retells. She really isn't even telling them to me. She's just telling them. But these stories are different from the ones about El Niño and her family. The stories about El Niño change. These stories she tells immaculately, speaks them as though she is praying, and no detail ever changes. Each story is always the same, like the stations of the cross or the beads of a rosary.

In the hall beyond my mother's room, a woman shouts at her husband, "If you're going to keep pissing around, I'll just go without you. Come on now, dammit." Neither, of course, is going anywhere.

The disturbance interrupts my mother.

"I have to get one of those things for the back door and then learn how to use it. Keep it locked," she says with some authority.

Another woman works her way slowly past the arguing couple. She wears a blue porkpie hat with the front brim up and the rear down. She looks like a mariner, face into the wind, feet bolted to the deck. She ignores the other woman's anger, leans a little further into the wind and slips beyond the storm.

Mother sends me for ice water. I return with a brown plastic cupful. I kneel beside her and hand her the glass. She takes a mouthful, then turns toward me with a twinkle in her eye. I know what is coming. I raise my hand, but too late. She spits half of her mouthful of water into my face. I take a towel from beside her sink and wipe my face dry.

"Ha," she says. "Got you."

And then she leans back and her chin falls to her chest. She is done with her defense. Her roommate, a small, bowed woman who once taught grade school in Mississippi, has crawled very slowly and painfully onto her bed where her cat, Cocoa, is sleeping. She strokes the cat slowly.

I lift the cup from my mother's hand, noticing, without meaning to, how polished and fragile, how old, her wrinkled hands seem. As her hand drops from the cup, she awakens. She looks up, directly into my eyes.

"The things I've heard," she says to me, "have ruined my mind."

And then her head settles back onto her chest. I place the plastic cup beside the porcelain sink. She dozes. I leave.

Most people would simply say that my mother is crazy. And they wouldn't be entirely wrong. But neither would they be entirely right. She *is* uninhibited. And she *has* unhooked herself from the moorings that keep most of us near the dock. But she isn't exactly crazy. That will come later.

In the meantime my mother tells stories. Over and over. And she tells her stories like prayers, because the stories are all that she has left. Inside of those stories is my mother, the real one. If she forgets the stories or she forgets to tell them, she will vanish. If she changes so much as one word, a piece of her will disappear. And when that happens, there is no way to get that piece back. When enough pieces are lost, what remains will, like a house whose single great timber has rotted through, collapse.

My mother's mind knows that. Because of that my mother is telling her stories, and she is deadly serious about it. Her mind insists that she do this. Her mind has noticed that pieces are disappearing. Her mind has backed itself onto the ropes and is counterpunching furiously. She is frightened and is fighting back the only way she knows how.

Unfettered by the pretense that slows so many of us, my mother has shown us mind in its purest state—a state where thought and defense are inseparable, a state where self has stared so long at four walls, one window, and a woman who was once a schoolteacher that it no longer needs pretense. Only defense.

"Shorty tried to kiss me." A posture. *"I had to carry all of the money down that dark alley."* Glancing over her shoulder, she has noticed me and all the others who seem to be gaining on her. For all that she ever was, she is gathering herself for one last battle. As she gathers near the bed in this little room, I envy her for her clarity of purpose and I wish that I, like her, didn't have to write about what is happening here.

———

Mind and neurons and lymphocytes and macrophages and neutrophils and natural killer cells. Defense. Self-defense.

The human immune system is the most nearly perfect defense on this Earth. For millions of years our immune systems have handled everything the fey god of evolution could throw at them. Infectious diseases too many to even mention were met and deposed. And before I was even born, my own immune system was drawing on its glorious history to assemble forces beyond imagining for my defense. Three or four months before my birth, stem cells (cells not yet fully committed to a single pathway of differentiation) moved out of the marrow in my newly formed ribs and femurs and found their way to my thymus. There, those stem cells became lymphocytes, T lymphocytes, and in the process of becoming lymphocytes my stem cells did a most unusual thing. Somehow, as the stem cells arrived, other cells inside my thymus sent a message to the stem cells telling them to begin rearranging their DNA. *Rearranging* their DNA.

DNA is strung out on chromosomes like words in sentences. And the stories those sentences tell are sacred. Change even a single word, and the sense of the story is usually lost. That's what mutations do, change the words in those stories. Because of that, many, even most, mutations are harmful, often fatal. When the words change, our chromosomes speak the wrong stories and a person is born without an arm or an eye or a chin, without enough insulin or growth hormone or hemoglobin. These altered stories are often so bad that we have evolved elaborate mechanisms to protect our chromosomes and the stories they hold—things like repair enzymes that remove altered DNA and molecules that slow cell division so that DNA repair can take place, complex pathways to destroy cells where too many changes have accumulated in the DNA, and finally we have hidden the stories inside of strong nuclear membranes.

Because our chromosomes are so protected, nearly all of the DNA inside of our cells reads just like the DNA inside of our parents' cells. Of course, we received some of our DNA from our mothers and some from our fathers. And some of our father's DNA moved from one of his chromosomes to another in the meiotic cell divisions that occurred during spermatogenesis. And some of the DNA in our mother's chromosomes performed a similar dance during oogenesis. So our chromosomes don't look just like those of either of our parents. But the chromosomes we got from our fathers contain DNA that reads pretty much just like it did in our fathers, and the chromosomes we got from our mothers have DNA in a sequence that reads pretty much just like it did in our mothers.

Except for the chromosomes inside of our lymphocytes. The chromosomes there, deep in the heart of our immune systems, are ours and ours alone. Some of the chromosomes inside of our lymphocytes—the cells that defend us from all of the microscopic threats in this world—look like no chromosomes found in our parents. In these chromosomes, bits of DNA are missing, bits of DNA have been added, and some bits of DNA have been moved from one place to another. All of that, all of those changes in the DNA of our lymphocytes, is there because of what happens in bone marrow and in thymuses. And much of that happens before we are born.

The gene rearrangements that occur in the DNA of stem cells while these cells are in the thymus occur in very specific regions of the chromosomes, regions that contain genes that encode proteins called T cell receptors. T cell receptors are the molecules that allow T lymphocytes to see the things that threaten us. When a stem cell arrives in the thymus, it is blind. When it leaves the thymus it has eyes for only one thing—the microbes that would have us for lunch if they could. Because of genetic rearrangement, each of us can make 10^{12} or more individual T lymphocytes with receptors that can distinguish among 10^{12} or more things that threaten us, and they can do that starting with only a few hundred genes. To get from a few hundred genes to a thousand billion different T cells requires a little sleight of hand, requires, in fact, a quick facedown shuffle of the genetic deck. And even after that shuffle a little cheating is required to deal all the different hands that will be required to sustain us. When lymphocytes are done shuffling their small decks, they toss out a little DNA here and they add a little DNA there until the hands that are dealt can be played 10^{12} different ways.

That's nearly miraculous. But it's dangerous as well. The whole process occurs randomly—like shuffling a deck of cards. Random processes don't have sense enough to guess at what the outcome should be. So random rearrangement of T cell receptor genes inside the thymus is equally likely to produce: T lymphocytes that are useless, because their T cell receptors don't do anything at all; T cells that could kill us, because their T cell receptors are specific for things like kidneys or pancreas or cartilage; and T cells that could save us, because their T cell receptors are specific for the virus that causes mumps or polio or influenza.

If we are to survive, something must be done about the useless and the dangerous cells. They must be destroyed. The thymus takes care of that, too. The thymus gets rid of the useless cells simply by looking, after rearrangement, to see which cells have T cell receptor molecules. T cells without receptors die. The thymus then begins to look for the dangerous cells. In this search, the thymus assumes that any cell that tries to destroy the thymus itself is dangerous. So the thymus gives all newly born T lymphocytes a chance to attack the tissues inside the thymus. Those T lymphocytes that accept the challenge and go after the thymus are quickly killed. So before it's done with them, the thymus has eliminated all of the new T lymphocytes that can't do anything and as many as possible of the new T lymphocytes that are dangerous to our selves.

This selection process has two interesting consequences. First, when our thymuses destroy those T lymphocytes that react with our own tissues, it opens holes in our defenses. Some of the molecules that lace together our thymuses are not very different from some of the molecules inside of some viruses and inside of some bacteria. When our thymuses destroy those T cells that would, if they could, destroy us, the thymus destroys T cells that we might need someday, opens doors we may later wish we could close. Viruses and bacteria know that, and over time they have evolved to take advantage of the doors that our thymuses leave open. A piece of HIV glycoprotein 41 looks like a piece of a human ribosome, a piece of another virus looks like human insulin, a piece of a third virus looks like human nucleoprotein. Keys that open doors intended only for self. Sometimes the results are disastrous; sometimes things go wrong and the viruses' similarities to our own molecules may cause our immune systems to turn on our selves. Keys to amazingly complex locks. Ways in.

Chinks in our armor, cracks in our walls. In fact, after the thymus has done its job, there is room inside of us even for other people. The molecules the thymus uses to test new T cells with are a group of proteins called major histocompatibility complex (MHC) class I and class II molecules. T lymphocytes can only "see" the things that threaten us after those things have been chopped up into little pieces and those pieces stuck to an MHC class I or II molecule. So that's how our thymuses offer the bait to newly born T lymphocytes, on MHC class I and II molecules. Immunologists named it the "major histocompatibility complex," because the first thing that we knew about these proteins was that when two people shared the same, or even mostly the same, major histocompatibility complex molecules they were much less likely to reject each other's kidneys, or hearts, or livers. Immunologists call two people who don't reject each other's kidneys "histocompatible." *Histo-* comes from the Latin word for "tissue." And the proteins that seemed to be the *major* factor in determining which people were histocompatible were MHC class I and II molecules—thus, "major histocompatibility complex." The thymus assumes that any T lymphocyte that attacks an MHC class I or II molecule while that lymphocyte is inside the thymus is a threat to human survival. So the thymus destroys all of those cells. Because of that, we are much less likely to accidentally destroy our own kidneys, or livers, or hearts. If thymuses didn't do this, we wouldn't be here. But also because of what our thymuses do, our immune defenses no longer recognize as "other" kidneys, or livers, or hearts from people who have the same MHC class I and II molecules expressed inside of our thymuses. In effect, we have created a password that allows our own cells free movement inside of us. But anyone else who knows that password may enter as well. Because of that, because of what our thymuses did while we were still in our mothers' wombs, parts of other people may pass through us unnoticed.

Our thymuses open holes in our defenses, but without our thymuses we die. Humans with DiGeorge syndrome are born without most or all of their thymuses. People with DiGeorge syndrome develop oral yeast infections, chronic pneumonias, skin rashes, and diarrheas, and these people often die shortly after birth. Mice born without thymuses will accept skin grafts from chickens and grow feathers. Without thymuses, we are not our selves.

But neither are our thymuses perfect at what they do. T lymphocytes that attack our thymuses are killed. T lymphocytes that don't attack our thymuses are released into the blood to create and maintain our defense, our self-defense. That makes pretty good sense. Clearly any T lymphocyte that went after a thymus would do a lot of damage if released into the blood. After all, cells of our thymuses are 95, maybe 99, percent like all the rest of the cells in our bodies. Most human cells do a lot of the same things, and most human cells do those same things using the same molecules. A liver cell is biochemically a lot like a thymus cell. So maybe 99 percent of those T lymphocytes that might destroy our livers are removed in the thymus.

But a liver cell isn't *exactly* like a thymus cell. Neither is a lung cell or a pancreas cell or a kidney cell or a blood cell. . . . And that's where the problem begins. From the outset, we have T lymphocytes with a thirst for self roaming through our bodies. At first glance that would seem to be a very bad thing. But inside of most of us, these lymphocytes never become a problem. Somehow we suppress most or all of those T lymphocytes that might destroy us. That's good. Even in the face of a few million cells that seem destined to cut us to pieces, we thrive. So there must be more compromises, compromises struck outside of our thymuses. There must be more ways in which our defenses fall short of perfect. Because of those compromises, there must be more places where others can slip through our defenses.

And finally, there are all those other things that we live with all of the time, especially bacteria—10^{14} or more of them that crawl on our skin, that are layered an inch thick in places inside of our intestines, that surround every word we speak, that share every kiss we give or take. Somehow we must make room for all of them, too, because we need them.

The finest defense system ever devised, but from the beginning that defense is porous. And it must be porous. It must be riddled with holes. Because when the holes close, there is disease—a most awful disease called autoimmunity. When there is no space inside of our immune systems for our selves, when we cannot tell the difference between us and not-us, we lash out at everything at once. Knees are crushed, eyes are blinded, nerves are shredded.

So, if we are to live with ourselves we must open holes in our defense. Holes that make room for our selves. But holes, too, that make room for others.

———

It is 1998. Wednesday, nearly noon. I'm waiting to order lunch at McDonald's. Behind the counter, people in purple uniforms are handing out hamburgers. In the line next to me, two young men, in their twenties perhaps, are trying to converse. One of the two isn't holding up his end of the conversation. It's not obvious why this man has so few words inside of him. His feet are pointed inward, his hands are curled back at the wrist, and his lips are thickened. He is different, but not in a way I recognize. The speaker works methodically with the slower man, trying to extract his lunch order. I admire his patience. His friend shifts his weight mechanically and speaks in monosyllables. Politely, I pretend to ignore them.

"May I take your order, please?" the woman behind the counter asks me.

"A fish sandwich, fries, and water, please," I respond, after the moment I need to disengage from what is happening next to me.

"For here or to go?"

"Uh, here."

She places a tray in front of me and moves off to harvest the food. I take hold of the tray with both hands and lean against the counter to study the overhead menu. A posture I never assume, but right now it allows me to avoid the two beside me. There are no fish sandwiches prepared. The woman waits at the stainless steel spot where sandwiches appear. I grip my tray and stare at the words that hang from the ceiling. Next to me, the conversation has stopped.

The slow man is now directly beside me. I continue to pretend to ignore him. He chooses not to ignore me. I stare straight ahead with nothing but peripheral attention.

Abruptly the man next to me reaches over and takes hold of my left hand. His hand is cool, dry, certain. He squeezes my hand softly but firmly, looks straight into my eyes and smiles.

His eyes are brown, his chin firm, his temples pulsing. He squeezes my hand once more and I am uncoupled from my past. I am, for an instant, an orphan. I reach to lay my right hand over his.

His friend notices what is going on between us. Before I can reach the slow man's hand, his friend pulls it away from mine and pushes it to the man's side.

"Keep your hands to yourself, Jack," he says, and then to me, "I'm sorry, sir."

"It's . . . no problem," I mumble, too surprised to ask for what I really want—his hand again in mine, if only for a moment.

His friend seems unaffected by the reprimand and continues to smile at me. Then the two men are served their meals and step away from the counter. I am too embarrassed to follow.

My past returns. I am served my food and sent away.

———

The human brain is one of the most nearly perfect neurological machines ever devised. It has given us words. It has given us vaccines. It has given us computers and quantum mechanics, bridges, locomotives, libraries, and nuclear weapons. It has given us prayer. It has given us poetry. And it, too, has dealt with nearly everything that this world could throw at it for the past million years or so. And all of that is done with words. Words and sentences, paragraphs and stories that spin inside our heads and out of our mouths. Some of those stories are just like the ones our parents tell. Stories of births and deaths. But inside of other stories, each of us has rearranged all of the words and added words our parents never spoke. Those stories are ours and ours alone. Much of the time our words speak of defense. Stories about who we are, and why we are here, and who we can trust. Some of the words we speak, we speak to our immune systems. Some of the aggression inside of our immune systems speaks to our minds.

Some diseases of the mind resemble diseases of the immune system so strongly that they are nearly indistinguishable. Schizophrenia is a disease often accompanied by paranoia—where a mind believes that everything threatens it, believes that all of those around it wish it harm. Autoimmunity occurs when the immune system imagines our very selves are a threat. And both diseases—schizophrenia with its delusions and hallucinations, its catatonia and emotional flattening, and autoimmunity with its crippling inflammations and sickening attacks on our own bodies—may be precipitated by infectious diseases. Many autoimmune diseases appear to involve infections, usually viral infections. Exposure of the mother and the fetus to influenza virus during the second trimester of pregnancy appears to increase the likelihood of later schizophrenia in the child.

Autoimmunity and schizophrenia, two diseases which blur self and nonself. Panic disorder and generalized anxiety disorder are psychological diseases where innocuous stimuli, like small spaces or simply the outdoors, are perceived by persons as life-threatening. Allergies are immunological diseases where innocuous stimuli, like pollen or peanuts, trigger life-threatening responses. Post-traumatic stress disorder—full of nightmares and flashbacks and thoughts that won't go away even for a moment—is a disease like multiple sclerosis that may be triggered by an infectious insult the immune system won't forget and because of that won't ever stop thrashing at the slips of myelin that surround our nerves. And in post-traumatic stress disorder, the levels of cortisol and serotonin, norepinephrine, and epinephrine are altered just as cortisol levels and epinephrine and norepinephrine sensitivity are altered in rheumatoid arthritis—another crippling autoimmune disorder.

In all of these diseases—immune or nervous system diseases—changes in our minds cause changes in our bodies. What we perceive changes who we are. Who we are changes what we perceive. Alterations in the HPA or HPAI axis close up the cracks in our walls, the holes in our defenses so tightly there is no place left even for our selves inside of the fortresses we have built.

Our thymuses and our brains create spaces inside of us, spaces for our selves. If we are to thrive, the spaces must remain. But others may use those spaces, too. Others who look, to our nervous and immune systems, like us. And still others, some of whom may help and some of whom may threaten us—viruses, bacteria, people in chain restaurants with cool dry hands, crazy women in rocking chairs losing their words, men on wooden benches with their own spit stuck to their stubbled chins.

Then there is no nervous system anymore, no immune system. Then there is only one thing—one thing that includes what we once called immunity, what we once called perception, what we once called self, what we once called other, and all that we once called mind—a thing we have no word for.

———

For a moment longer, I sit on the wooden park bench and watch the backs of people passing me by—his words ringing in my ears, his spit still soaking in the acrid dust. Then suddenly, I am on my feet, after him. He can't have gotten too far in the few minutes that have passed since he walked off. But there are so many people, it's like swimming upstream. No one will make room for me. I push against men and women, trying to punch a hole in the crowd. People are startled, angry. Some do their best to slow me. Finally I see him, a block ahead. I push harder. A hole opens and I spurt forward. Then slow. I don't want to frighten him into one of the doorways that line the walk. I continue, just fast enough to overtake him.

I reach for his tattered sleeve. He stops and turns.

"Isn't there something I can do? A meal? A place for the night?" I ask him.

"No," he says, looking nervously at the people passing on either side.

"I want to do something," I say.

"Leave me alone," he says and tries to twist free. But I hold him and turn his shoulder until he is looking at me.

"You were wrong, you know," I say to his bruised eyes.

He stares at me blankly.

"You were wrong when you said it was *my* idea from the start."

"What?"

"You were wrong," I said again. "This isn't my fault. None of it was my idea."

He draws back, then looks at me curiously for an instant.

"Who cares?" he finally mutters at me and looks away. "Who gives a shit?"

Then he pushes my hand from his shoulder and shoves his way into the crowd.

"I do," I say to his retreating back. No one notices.

Overhead the red-shifted light of a thousand suns screams past. And a flock of crows rises from the roof, then scatters like a handful of thrown gravel.

"I do," I say again. "I do."

It is a sad day when you find out that it's not accident or time or fortune but just yourself that kept things from you.

—LILLIAN HELLMAN

It's when we are given choice that we sit with the gods and design ourselves.

—DOROTHY GILMAN

Light and Shadow

It feels as I imagine a woman's kidney might—slithery in its capsule, firm, smelling of blood and female urine. I squeeze, the kidney oozes slightly. Curious, I raise my hand to my mouth. The taste is . . . salty, perhaps bloody.

Abruptly, my mother pushes me away.

"That roast is for tonight's dinner. Keep your hands and tongue off of it. In fact, as long as you're blind, I want you out of my kitchen. Out of my sight and into your room!"

"Mom, I'm only—"

"Go on. *Now!*"

I make my way to the kitchen door, feel for the jamb, ensure that the door is open, and step into the hall. Immediately, I stumble down an unseen step, grope wildly for the wooden banister, and bang my forehead against the coatrack. Finally my hand lands on the banister. I steady myself, feel for the lump on my forehead, then reach with my toes for the stairs—one at a time. Feeling for each, I make my way down and into the bedroom I share with my brother. He's away just now. I leave the light out, sit on the bed, and squinch my eyelids a little tighter.

I'm completely blind.

I wasn't born this way, though. In fact, I could see fine until this morning.

———

Last night at Bountiful's only theater, the Queen, I saw *Them*—a
movie about giant, ravenous ants. Ants created by atomic fallout.
Ants with a vision of world domination. Ants with pincers big as
chain saws. Lots of gore. The movie scared the bejeebers out of
me. All night long, I was pursued by venomous dreams, and it
was one of those dreams that blinded me.

In the scene that played itself out inside of my head, a group
of soldiers had cornered an especially nasty ant. I knew from our
backyard that ants could sting. I knew from school that their
sting was the bite of formic acid. How it itched and burned. In
desperation, one of the soldiers tried to toast the mutant ant
using a flamethrower. In the process of the crisping, the ant—
with good cause, I guess—spritzed the soldier with formic acid.
As the pH of the soldier's face reached about 0.4, his corneas
turned the color of buttermilk. The camera inside my head
panned slowly back; the soldier stared slack-jawed into the darkness where an instant ago there was a world of color. Then he
screamed a scream from the black heart of terror and I awoke
in a paroxysm of fear, rattling my brother's bunk above me. My
brother cursed me briefly, and then went back to sleep.

Only ten years old, I had never imagined anything so horrible.
And as if that weren't enough, as I lay there in the dark, I imagined the soldier groping his way through the rest of his life with
bandages covering his curdled eyes, his friends watching sadly
as he tries to button his uniform, shrugging their shoulders help-
lessly as he bumps oafishly against tables and walls, holding his
hand to guide a forkful of peas to his mouth. And then I imagined
I was the soldier.

A friend of mine, Tim Kelly, had lost one of his eyes to an
arrow fired in fun. The suction cup gone, the wooden shaft emptied his eye like a spoiled grape. That could have been me. In
my dreams it was me.

Giant ants tromping through my head, I lay awake the rest of the night thinking about blindness and what it would be like to live without eyes. Was it terrifying? Did it change everything? Did it hurt?

Near morning, I decided that before blindness was forced on me by some horrible accident involving an arrow or an ant (which I was certain was out there somewhere waiting for me), I needed to know. I needed to know how blindness felt.

I would spend a day in darkness. I would go and see what blind people see.

———

As I sit here on my bed, my day of blindness is nearly done. The sun has set. I can tell the darkness around me now is real—deep and angry. The air smells like snakes and there is a ringing in my ears.

———

I believed at the end of that day, a day spent mostly in darkened rooms with my eyes squeezed tightly shut, that I knew how it felt to be blind. Of course, I didn't. My day of darkness was no more than that. Today as I write this, I believe that I know what it means to see. But of course I don't.

In his story "The Spiral," Italo Calvino reimagines the origin of the things we call eyes. "I had conceived an idea of my own," the protagonist, Qfwfq, states, "namely that the important thing was to form some visual images and the eyes would come later in consequence." Calvino's eyes arose, not as we have been led to believe, from the accidental photophilia of some errant neuron—once meant for the taste of salt or the sound of waves against rocks. Instead he proposed that sight arose in response to simple beauty: a turbinate shell in coral and aqua, the curious curl of the nautilus with its brown sunbursts, the essential intricacy of the primal helix. Beauty itself and the need for a vision of that beauty that drew, like a crocus, an eye through the skull of the blind.

For Calvino, eyes were windows that shot their bolts and rose in their casements at the first touch of beauty. A remarkable thought. And surely there is beauty in the birth of an eye, any eye. And just as surely, none of us will ever know for certain whether it was beauty's or some other's hand that first pierced the clay and pulled from it an eye.

But eyes aren't windows. No matter how quickly and easily that image leaps to mind, eyes are not openings, cut like glass into the bone of our darkness. Corneas, perhaps, are windows. But windows only for the tiny darknesses inside of eyes. The sunlight that falls through corneas never makes it inside our bodies.

Neither is the image painted on our brains a windowpane's image of what stands beyond the wall. The light that enters eyes passes first through the cornea, then through the aqueous humor (the liquid below the cornea), then through the lens of the eye, then through the vitreous humor (the liquid that fills the eyeball itself), and finally focuses onto the retina. Retinas are thin sheets of photosensitive cells that line the backs of eyeballs. Then, since our retinas seem to be installed backwards, the light must pass through a layer of ganglion cells, two plexiform layers of cells, and the inner and outer layers of nuclear cells, before it finally reaches the rods and the cones (the optically active cells of the eye) at the innermost layers of the retina. There, what little light remains is extinguished. After that, it gets complicated.

The few photons that reach the rods and cones are absorbed there and converted into electrons. Via a series of neurons, those electrons are then piped to the brain. Most neurons don't work like this. Most neurons transfer signals via action potentials—waves of depolarization that result from transfer of sodium and potassium ions across the neuronal cell membranes—not transport of electrons. Optic nerves transfer electrons—just like the copper wires inside the walls of our homes—directly into our brains. And that's good. Because direct transfer of electrons allows for a graduated response in direct proportion to the amount of light that enters the eye, something that would be impossible using action potentials as other nerves do. The graded responses of optic nerves allow us to perceive evenings and mornings, not just noons and midnights. Regardless, what arrives in our brains are electrons, not photons. The photons burned their final flames in the backs of our eyeballs. Before we've even noticed, the light is gone.

Worse yet, along the way other things are done to the electrical signals sent by our rods and cones, things that heighten contrast and enhance or add colors. Even the blind spots—created by the optical discs at the backs of our eyes—are filled in with things that our brains imagine should be there.

So by the time an image finally flickers inside our skulls, that image is unlike any fallen from a window and bears who-knows-what sort of resemblance to the things that might actually be happening before our eyes.

To further complicate matters, retinas aren't composed of single sheets of light-sensitive cells. Retinas are split into two sheets. One of those two sheets lines only half of the rear hemisphere of an eyeball. So in each eye, there is a temporal half-retina (nearest the temple) and a nasal half-retina (nearest the nose). Each of the half-retinas is attached to a fiber of the optic nerve. As these nerves move back through the eye sockets, the fibers from the two nasal half-retinas cross over one another. As a result, the left half of my visual cortex receives signals from my left temporal half-retina and my *right* nasal half-retina. While the right half of my visual cortex must work with images from my right temporal half-retina and my *left* nasal half-retina.

So our retinas transmit at least four discrete images to each of our brains. And where two of these images fall upon our brains has nothing to do with where things are actually located in the "real" world.

Once the visual signals arrive in our brains, things get truly complex. The electrons from the optic nerve are delivered first to two small structures called the lateral geniculate nuclei. There, individual neurons accept signals from one eye or the other, but not both eyes. From the lateral geniculate nuclei, the signal is exported via the optical radiations to the primary visual cortex at the back of the brain. And from there, visual information is transported to more than twenty-four other regions of the brain.

And then the holes are filled in. Where the optic nerves attach to each of our retinas, there is a hole—a space where the eye is insensitive to light because of the nerve itself. The result is a circular gap in the central part of the visual field in both eyes. We never see that gap, though, because our minds fill in the gap with what the brain imagines should be beyond the hole—*imagines* should be beyond the hole. Our brains makes a best guess at what probably is in the center of our fields of vision and then creates an image to fill that hole. A portion of the image we see isn't necessarily there at all.

Then the whole thing is processed. First of all attention, the area of our visual field where we focus, filters out the less important stuff. For example, while reading this page at least two things happen. Though you can see the whole page, the only part you see well is the word or few words you are currently focused on. Second, individual letters are generally not seen as critically as while proofreading; most reading gathers only whole words. Finally, we process and interpret what we see. It takes time for a vision to register in the mind. If we aren't given between 50 and 125 milliseconds between images flashed on screen our minds cannot distinguish them. During that 50 to 125 milliseconds, one of the things our brains are doing is trying to fit what we saw into a preconceived notion of how things ought to be—a mental paradigm, or maybe more accurately a metaphor. It takes time for the light falling from an elm into our eyes to assemble itself into a tree, into the word "tree." A while for the odd image created by rolling film to assemble itself into what seems to be motion, a while longer for the spots of a pointillist painting to assemble themselves into the Eiffel Tower or a woman with a parasol.

The electrons gathered from the optic nerves are sorted, processed, interpreted, and reassigned before we ever "see" anything. Then, from the mismatched mosaic spread across most of the involuted geography of our brains, our minds construct rather than receive pictures—Aunt Helen, beauty, pleasure or revulsion, invitation or threat, lunar halos, and the constellation Gemini. We *construct* pictures.

————

In *An Anthropologist on Mars,* Oliver Sacks describes a patient given the gift of sight as a grown man.

Finally in mid-September, the day of the surgery came. Virgil's right eye had its cataract removed, and a new lens implant was inserted; then the eye was bandaged, as is customary, for twenty-four hours of recovery. The following day, the bandage was removed, and Virgil's eye was finally exposed, without cover, to the world. The moment of truth had finally come.

Or had it? The truth of the matter (as I pieced it together later), if less miraculous than Amy's journal suggested, was infinitely stranger. The dramatic moment stayed vacant, grew longer, sagged. No cry ("I can see!") burst from Virgil's lips. He seemed to be staring blankly, bewildered without focus-

ing, at the surgeon, who stood before him still holding the
bandages. Only when the surgeon spoke—saying "Well?"—
did a look of recognition cross Virgil's face.

Virgil told me later that in his first moment he had no idea
what he was seeing. There was light, there was movement,
there was a color all mixed up, all meaningless, a blur. Then
out of the blur came a voice that said "Well?" Then, and only
then, he said, did he finally understand that this chaos of light
and shadow was a face—indeed the face of his surgeon.

Virgil, like other people given sight as adults, had never *learned*
to see, never learned to construct pictures inside his brain. And
many of the things we take for granted—the differences between
photographs and reality, the difference between looking at a window and looking through a window, where the window ends and
the image begins—Virgil, with his newfound sight, never mas-
tered. And for Virgil, like many others blind into adulthood, the
initial euphoria of sight gave way to a severe depression over the
difficulties of adapting to a new sense. The gift of sight became
a photovoltaic curse.

None of us, it turns out, immediately "sees" the signals that
arrive in our brain. First, what we have "seen," as I said, must
be processed. That takes as much as 125 milliseconds. Not long.
But long enough for several interesting things to happen. First,
we focus on particular aspects of the scene and eliminate other
"unimportant" aspects of our vision. Pieces of what we have seen
actually disappear. Then we interpret what we have seen, and
look for an explanation that fits our expectations. This inter-
pretation is most apparent in optical illusions—a series of lines
that may look like a cube on the flat piece of paper, a vase that
becomes two people facing one another, a letter or two that
aren't there as we scan a page of print. We search for simple

explanations of what we have seen, and we force our visions into the clothing of those explanations. Vision as narrative. Vision as chimera.

An especially graphic demonstration of this occurs in people who have had their corpora callosa severed, a procedure sometimes used to treat certain types of epilepsy. The corpus callosum is a bundle of nerve fibers that allows the two sides of a person's brain to communicate with one another. When it is severed, the two halves of the brain operate in isolation. Normally, it is the left half of a human brain that is most directly associated with speech, and it is the right half that is most directly associated with vision. If a person with a severed corpus callosum is shown a scene in such a way that the image enters only his right eye and then he is asked to point to the image and identify related objects by touch with his left hand, he can do that. But if he is then asked to explain what his left hand was doing, the left (language) half of his brain will *make up* a story to explain what his left hand did, a story that includes a vision that his left brain never saw. Those who conduct this kind of research call this *confabulation.* Those of us less familiar with the language of psychology might call it *lying.*

Visions are *processed,* and in those few milliseconds, things change.

At birth, then, we are given eyes, not vision. Vision we learn. As children, it's simple learning to cope with the assault of imagery, but learning to see is nearly impossible for adults. Vision isn't a snapshot of what is "out there." Vision is an intricate process through which we *construct* the world around us. It is easiest for us to learn to do that as children. Just as there is with speech and immunity, there is a window in childhood when we learn to see easily. Later, learning to see is nearly impossible. Considering the complexity of the images presented to our brains, and the intricacy of the pathways by which these images reach our brains, perhaps it is understandable why those who are blind into adulthood find it so difficult to learn to see.

But what does that tell us about seeing itself? Among those of us who learn to see as children our visions more or less agree. But we learn that. We learn to say that we see the same things— just as we learn to say: "That is the call of a whippoorwill," or "I smell bacon frying," or "I know how you felt." But because we all agree to say that we have seen a whale doesn't mean we have all seen the same whale. Each of us learns in his or her own way to see. We learn to take the electricity that our eyes place before our brains and to make from this a world, a world of light and color, a world of depth and meaning. We *learn* to do that.

I have a bit of a knack for multiplication. I can multiply numbers like 65 × 225 in my head. I do it by reducing the numbers to something I can work with:

$$65 \times 225 =$$
$$((2 \times 225)3 \times 10) + ((2 \times 225) \times 2) + 225 =$$
$$13{,}500 + 900 + 225 =$$
$$13{,}500 + 1125 =$$
$$14{,}625.$$

I don't know of anyone else who does it like that, but we all agree the answer is 14,625. I *learned* to do that, even though no one taught me how.

When we learn something, we take a bit of what is outside our bodies and we make it into something of our own, something personal. We learn, that way: the taste of blackberries, the smell of human skin, Kant's Categorical Imperatives, how to hit a baseball. But no two of us learn any of it exactly the same way. Ask anyone about hitting a baseball, ask anyone about the relevance of Kant's Categorical Imperatives.

We learn to *look* as well—to look at a night sky, a woman's calf, a man's wrist, a tree. And if we learn to see in the same way we learn other things, then none of these things appears alike to any two of us—no two of us have ever seen exactly the same thing. The same Renoir, the same Denali, the same sunrise. A single human blindness may extinguish an image that no one else can ever reconstruct.

Eyes aren't windows. Eyes are something else, something we have no word for. There is no synonym for eye.

———

But there *are* windows, more than any of us ever dared imagine.

The pattern of the sun is tightly woven into the flesh of humans and other animals.

As long as we are not kept from the sun and moon, we wake and sleep in twenty-four-hour cycles. Our hearts' and brains' rhythms rise and fall in twenty-four-hour cycles. Our immune systems strengthen and weaken as the sun rises and sets. Our bodies warm and cool to the rotation of the Earth. Our hormones surge and ebb as the planet spins. We are creatures of the sun.

And our contracts with that sun are kept by the clocks we are born with, the clocks inside. All of us—from slugs to bats, from morning glories to moose—move to circadian rhythms—twenty-four-hour cycles that control everything from the production of sperm to the wave patterns in our brains. The word "circadian" comes from the Latin *circa* and *dies*—"about a day." These solar rhythms are kept in us by circadian oscillators, better known as biological clocks—poorly understood sites in our bodies where we produce the chemicals that tie us to the sun, where our cells sing the star songs that bind us.

In insects and many other land-dwelling species, including mammals, six genes—*clock, cycle, timeless, period, double-time,* and *cryptochrome*—and the proteins these genes produce are largely responsible for the accuracy of biological clocks. The levels of proteins produced by these genes rise and fall with the sun. That is (and this is the crux), sunlight itself activates the cryptochrome protein. Cryptochrome then sequesters and inactivates the time protein. That inactivation allows clock and cycle proteins to initiate synthesis of more period and time proteins throughout the day. In the dark of night, then, these proteins move into the nucleus. This allows for production of new proteins—like the hormones that regulate our biological cycles—and inhibition of clock and cycle proteins, until the sun rises, cryptochrome is again activated, and the whole process begins again. A twenty-four-hour routine, driven by sunlight.

One place in the mammalian body where all six of these genes are known to operate is inside the suprachiasmatic nucleus, a small group of cells that lies just above the optic chiasm, just above the place our temporal optic nerves cross over one another in our brains.

Because of that—the evident participation of the brain in circadian rhythms—for a long time scientists imagined that all biological clocks must be in the brain or at least controlled by the brain. Scientists also imagined that the brain was the only human organ hard-wired to cells that could "see" sunlight. So the brain must be responsible. Right? How else could we possibly keep track of the sun? How else indeed.

It now appears that our bodies are filled with clocks, and most of them have nothing to do with our brains or our eyes. Chemical maps of *timeless* and *period* have revealed clocks in arms and legs, clocks in toes, clocks in livers and kidneys, clocks in testes, clocks behind knees, clocks in lungs, and clocks in hearts.

In front of each of these clocks there is a window—a window to let in the sun. There must be, because it is the light of the sun that turns the hands of these clocks.

And though most studies have been done in fruit flies and mice, it appears humans are no different. We are filled with windows. We are full of sunlight.

Whether eyes are windows or not, we are as luminous as the great cathedral at Chartres. We are votives at the altar of a star.

———

There is a certain darkness, too.

Inside the pigments that coat the roof of the Sistine Chapel, Michelangelo Buonarroti has spread, like a storm, the immense story of Genesis. The Creation of Adam is the best-known part of this story, with God lazily stretching His right forefinger toward Adam's left, a soul about to flare inside the first human. But there is also a panel depicting the creation of light and darkness. In the center is God, wringing light from the clouded heavens—the darkness spills down His left side, the light kindles along His right. God's robe is the color of salmon, His beard a storm's cloud, and His dark eyes seem only distantly involved with what He is doing. There are men in the fresco as well. Most appear to have little interest in God's new light. But to God's left, a man is seated on a stone bench. This man is bent backwards and stares openmouthed at the interplay of light and dark. I am drawn to that man. There is no way to tell from the painting whether Michelangelo imagined this man bent by the promise or the horror of what he has seen, whether he is overcome by the joy of what he has just been given or the deathly fear that it might someday be taken from him.

In spite of that, my attraction to this man, standing beneath that fresco it is easy to ignore all of the men in the painting and to look only upon God's light. But on that day, God created both light and darkness. The darkness is God's child, too.

———

If you peer into someone else's eyes, stare at the space beyond their pupils, all you see is darkness, an oily blackness. The color, I've imagined, of the darkness inside each of us.

Of course, it isn't the color of anything at all that I've seen there. It is a tiny darkness, nothing more.

What we see when we stare through another person's pupils is her retinas. Retinas have evolved to absorb nearly all of the light that passes the corneas. Because retinas soak up all the light, like small black holes, retinas are very dark. Not much light escapes them. So the blackness we see inside of others is only the darkness of the vitreous humor inside their eyes, not the tomb-light from within their bodies.

Looking out or peering in, eyes are not windows. But there is darkness.

Tonight I am naked, lying atop my sleeping bag at the edge of a very deep canyon. My brother Michael and his son John sleep nearby. The day's sun is still bleeding from the rocks. It was a hot sun. Overhead, there are a million, million stars—old light that pierces me now in a thousand places. To my eyes, for my life's time, no one of those stars will ever appear to move. The Great Bear, Cygnus, the Northern Cross will be just as recognizable to my grandchildren as they were to my grandfather. Constant, reassuring. But contrary to my perceptions, the stars and the galaxies are in motion, and using powerful telescopes, cosmologists have mapped this motion.

Those maps reveal a cosmic mystery. Observable matter—planets, stars, galaxies—account for only about 10 percent of the observable gravity, the force that holds all of this together. To fill the void, cosmologists created "dark matter"—matter invisible to us because it emits and reflects no light, *dark* matter. And though some have argued against it, there is good evidence that dark matter actually exists. Unexpected perturbations in the motions of the stars and, especially, gravitational lensing—the bending of light by massive unseen objects—seem inexplicable without the gravity of dark matter.

Dark matter may be undetected massive particles, dense but hidden stellar objects, both, or neither. But it appears that more than 90 percent of all the matter in the universe must be dark matter. That is a great deal of darkness. A great deal that our eyes, no matter where we look, can never see. But the evidence stands. Among all that is lighted there is much that is darkened.

We are full of windows. We are full of darkness.

We are, each of us, as dark as Ramses' tomb. We are the black ink that fills the period at the end of the universe. We are cups of cosmic shadow.

The air tonight smells of scorpions and screech owls. I close my eyes tightly, the darkness intensifies only slightly. Blood rushes in my ears, my mouth fills with saliva. I open my eyes, the light burns only a little more brightly.

Tonight, I can't help but wonder at that—the distinction between seeing and not. Wonder about the stars blazing overhead and the massive darknesses in between. Wonder at what is lighted and what is blackened, what is revealed and what is hidden. Wonder, tonight, whether any of the planets spinning around any of those flaming stars is ruled by venomous black ants.

Watermarks

At its heart, pathology is the study of old water—how it once moved, why it is now still. In fact, all of biology is truly the science of water. Biologists are hydrologists. "The role of humans in this world," it has been joked (mostly by beer drinkers), "is to move water from one place to another." The only inaccuracy here is that the joke is limited to human beings. The role of *all living things* and most of the rest of this planet is to move water from one place to another.

In spite of that, in spite of our intimate and irrevocable relationship with the wetness of this world, it is rare that any one of us imagines that the water we move is a part of us. Instead, we imagine that the water is only passing through us. Water is like the darkness in van Gogh's *Starry Night*—necessary, because you couldn't see the stars without the darkness, but not a part of the stars themselves. We *drink* a glass of water. We *make* water off the back stoop on a winter night and watch while our urine steams on the cold ground. But we are not the water. This isn't a glass of me (or anyone else for that matter) that I am about to drink. This isn't some part of me puddled there on the rock-hard ground. It's only water.

———

It is winter, 1949. I am three years old, living with my family in a small house on MacArthur Road in Coffeyville, Kansas. I have

just awakened from my nap. At the foot of the bed a black Bake-
lite vaporizer is boiling water in a crusty aluminum cup. My
mother set the vaporizer there to fill the air with steam and ease
my congested lungs. I watch the vapor for a moment, watch it
rise in little jets like fairies' breath on a winter night. But the
steam doesn't hold my attention for long. Little does. I am
quickly bored with the vaporizer and long for grander vistas. My
mother is sleeping next to me. So, to get out of the small bed, I
have to crawl past her feet without waking her. As I am trying
to do just that, my foot snarls in the electric cord of the old
vaporizer. I struggle to free myself, trying not to wake my mother,
and I pull the whole boiling thing into my lap. There are maybe
three cups of water still steaming in the small pot. Not much
water by anyone's standards, but enough.

As my father carries me through the glass doors of the hospital,
I lose consciousness. He thinks water has killed me. Instead, I
spend nine weeks in the hospital absorbing new water through
steel needles stuck in my chest—lots of attention, lots of bandages. And then we mostly forget about it.

For most of my life, especially before and after gym classes, I
resented that—that thing the water did to me. The forgetting, I
resented that, too. In spite of my resentment, in spite of my
embarrassment, in 1966 the mark the water left on me kept me
from the rain forests and rice paddies of Vietnam.

————

Bernadette Soubirous was born in a small town in southern
France. When she was ten years old, a chip from a broken mill-
stone blinded her father, and that began a downward spiral for
her family, leaving them among the poorest of the poor. By the
time Bernadette was fourteen, her family was living in a one-
room house that had once been the jail, was once filled up with
drunks, petty thieves, and part-time whores. Every morning, so

the story goes, it was Bernadette's job to leave her jailhouse home and gather firewood. This February morning was cold, colder than usual for late winter so near the mountains. Bernadette didn't care much for gathering wood. It was hard work, especially in winter. But this morning, she was in luck. There was a pile of driftwood stuck at a nearby bend of the river Gave, and she could quickly fill her basket. But as Bernadette was crossing the river to get the wood, a sound like thunder rolled up out of the ground. Suddenly, in a grotto behind her—the grotto of Massabielle—there stood a beautiful woman wrapped in a golden cloud. In the woman's hand was a rosary. Bernadette fell to her knees, and together she and the woman-in-the-cloud prayed the rosary.

———

I have this recurrent dream. The dream takes several forms. In one iteration, I am sitting in my house, my wife is somewhere else, and I remember an apartment that I rented several years ago. The apartment was smallish, furnished with old things of mine—a burgundy sofa, a desk, some wooden shelving. And even though I have been living in this house with my wife for several years, I realize that I never gave up that apartment. I never gathered my things, paid the last month's rent, and moved out. A lot of my stuff is still there. I know that I left some books behind, though I can't recall which ones, several shirts, a couple of lamps I liked, a television. I think, as I sit in my dream, that I really should have kept paying the rent on the apartment. It seems like I might need it—the place, my things—again. And it makes me feel, not quite sad, but foolish, I guess, to have simply forgotten about it. The dream usually ends with me deciding I should go soon and see if the apartment and my things are still waiting for me, even though that seems improbable. I haven't paid the rent in years.

In another form of this dream, I'm living in a large house with a sunken living room, and I discover the house has other rooms I have forgotten about—rooms off halls and stairs that lead away from the sunken room; rooms that fascinate me, but that I just haven't gotten around to visiting in years; rooms with thick, plum-colored tapestries and heavy, oaken furniture stained dark; upstairs rooms with views into the gardens; breezy rooms with latticework roofs held up by spiral pillars the color of lead; rooms out back. Rooms that I am drawn to. Rooms that frighten me, fanciful rooms, and rooms I might never have imagined. I wonder, in my dream, why it has been so long since I was in those rooms. It is interesting, though, that this house—the house-of-rooms I find in my dreams—seems also to have never been *my* house, but a house I rented. As in the other dream, what it is that I have forgotten here was never mine, only leased to me.

Perhaps dreams are merely Kodachrome flashes that spin across old screens within our skulls. Useless. Of little consequence. But then again, maybe dreams are something more—old memories stirring in the water inside each of us. Visions left there by others to remind us of things we have forgotten.

———

During the following two weeks, the woman appeared to Bernadette eight more times. On her ninth visit, the woman made a request of the girl. "Drink," she said, "and bathe yourself in this fountain." Looking about, Bernadette could see no fountain. "Where?" Bernadette said. "Where should I bathe?" "Here," the woman said and pointed at the soil near her feet. "Dig here." And Bernadette began to dig. Soon a small trickle of water arose from the hole she was digging. Bernadette stopped scooping the soil and washed her hands, then drank from the spring. By next morning the flow from the spring was nearly a river. The water was cool and crisp as the sounds the bells made in the mountains overhead, as clear and complete as a thought. That was February 25, 1858. More than 140 years later, the spring still flows at Lourdes in southern France.

———

It is August. It is hot, and the steel beneath my bare thighs is urging me to jump. It's maybe 1963. I'm about eighteen. Along with three others, I have just walked to the top of a train trestle above a bridge over a reservoir somewhere in Idaho. The iron beam is black, approximately ten inches wide and approximately the temperature of the sun. Water is lapping gently against the bridge pylons sixty-five feet below where I sit. I want to skip this. And I would if I could skip what we have come here to do and go back down to the bridge below. I am that frightened. But I am more afraid of the walk, or crawl, back down the sloping, burning arms of this iron trestle.

So before the others jump, before I am left alone and terrified here above the bridge, I force myself, at last, to jump. Even in the air, even fifty-five feet above the water, I realize I have made a tactical error. When I launched myself from the bridge, I pushed too hard. And now I am rotating slowly forward. I flail with my arms in an effort to right myself, but it is too late. The water is rushing up at me at thirty-two feet per second per second.

I hit the water slightly facedown, but mostly parallel to its glossy surface. It feels like I have hit the hard green slab of a concrete tennis court. The air is slammed from my lungs, and my body is driven fifteen or twenty feet below the surface. As quickly as I can, I push upward. From below, the surface looks like the rough back of a poorly silvered mirror. My chest is flaming. I kick, kick again, and the world above water seems no closer. Kick, kick again. My chest insists that I draw a breath. And I nearly do.

Suddenly there is an explosion in the water next to me, and an arm the size of a small tree wraps around my chest and shoves me upward. My head breaks the surface, and I gasp as I haven't gasped since I rose from the water at birth, drawing bucketfuls of air into my screaming lungs. And then I start down again, swallowing and choking on the water.

My friend, Tim, who jumped behind me as soon as he saw what I had done, grabs me by the hair this time and yanks me to the surface again. I flail. He stays just far enough away from me that I can't drag him back down and just close enough to keep my head above water. Finally, I calm slightly and he pulls me across the six or so feet of open water that lies between me and the bridge pylon—a pylon I'm certain did not exist thirty seconds ago.

I grasp at the mossy concrete, slide below the surface once more. My friend lifts me again and guides my hands into the chipped and cracked spaces of the concrete. I float. He laughs. I cry.

———

Without water the night sky would burn much cooler. The condensation of cold clouds of gas into fiercely hot stars is dependent on water. Water is a remarkably efficient coolant, in large part because of the surprising number of vibrational and rotational states of water molecules. Because of that, because of water's facility as a coolant, it absorbs a large amount of energy during protostellar collapse, cools the birthing star, and accelerates its coalescence into a stellar furnace. Without water many fewer protostars would ever achieve the density necessary for ignition of the fusion reactions that light our nights. And of course, without stars there would be nothing else.

So indirectly, at least, water drives every act that makes this universe living, every act that makes us human. We are, after all, the stuff of stars. But water drives humanity in much more immediate ways as well. Proteins are the pieces of humans that move muscles, build cells, make sugar and fat into usable energy, turn genes on and off, write poems, carve santos, sing psalms, fuse sperm and egg into more human beings. Proteins are enzymes. Without proteins no human act could occur in real time. Proteins are cartilage and bone, muscles and toenails and hair and most of everything in between. Proteins do nothing without water.

To do what a protein was created to do, it must assume the proper shape. That is not nearly as simple as it might seem. Most proteins have the ability to assume many, often millions of, different shapes. But only one shape will allow a particular protein to do what it is supposed to do—to register the wavelength of a photon, record the vibration of a violin string, point a finger at a falling star. Many things aid proteins in finding the right shapes, but the most important of all is water.

Some parts of all proteins interact easily with water, some not so easily. The parts that interact easily are called hydrophilic, which means "water-loving." The parts that interact not so easily are called hydrophobic or water-fearing. Hydrophobic parts of proteins are greasy like old oil and don't mix well with water. When a substance like that is forced into water, it can cause water to assume very elaborate structures called clathrates. Clathrates look a little like birdcages or geodesic domes—with a molecule of water at each of the apexes of the dome—surrounding each hydrophobic molecule in water.

These clathrates are highly ordered structures. And yet, the natural tendency of this universe is toward greater disorder, toward greater randomness. Stars burn out, clocks run down, highways disintegrate, and buildings topple. That tendency is called entropy. An overall increase in entropy as part of any spontaneous reaction is pretty much demanded by the second law of thermodynamics. When water molecules are forced to assume shapes that look like birdcages, the water molecules become more ordered than they are in plain water. That seems to violate the second law. And seeming violations of the second law require the input of large amounts of energy.

Proteins are strings of molecules called amino acids. As proteins are made inside of cells, these strings spool off the ribosomes (places where proteins are made inside of cells) into the water-rich cytosol (the liquid that surrounds and fills the nucleus and other organelles of living cells). Some of these amino acids are hydrophilic and some are hydrophobic. Around the hydrophobic amino acids, the water inside the cell begins to organize itself into clathratelike structures. Entropy drops, and that can only happen when a lot of energy is added to the reaction.

Mathematically, the Gibbs Free Energy equation describes the energetics of any such reaction as $dG = dH - TdS$. In English, this equation says that the change in the Gibbs Free Energy of the system (dG) is equal to the change in enthalpy (dH, basically energy released or absorbed as heat) minus the temperature (T) in degrees Kelvin times the change in entropy (dS). A chemical or biological event will only happen spontaneously when the overall Gibbs Free Energy is a negative number, and the more negative the better. That means that if the available heat is low (usually anywhere beyond the fierce flame of the sun, and away from the fires at the center of the Earth), dS must be positive. If it isn't, the negative TdS term will be a large positive number, and that means that the only way that the reaction will ever occur is if a large amount of energy is pumped into the system from the sun or the center of the Earth.

So it is an energetically bad situation to be extruding proteins into the water of a cell with all of the proteins' hydrophobic amino acids exposed. Because of that, proteins can gain a lot of energy from their surroundings if they fold up in ways that bury their hydrophobic amino acids and expose only their hydrophilic acids, which interact with water in much more energetically favorable ways. So proteins, as they are born, fold like fists where the palms are full of hydrophobic molecules and the knuckles are filled with hydrophilic molecules. Then the exposed part of each protein is water-loving and the inside of the protein is water-fearing—but emptied of water. The energy gained in the process of folding is called hydrophobic bonding. If it weren't for hydrophobic bonding, there would be no human beings, or aspen trees, or lynxes, bison, bugs, mushrooms, or mosquitoes.

Water is also stuck to the outside of every protein inside a living thing. That water is called the water of hydration. Without the water of hydration, proteins again lose their shapes and can't do anything that resembles living.

Water also keeps all of the salts inside of us dissolved, participates in the hydrolysis of fats and proteins and sugars, and does a slew of other things as a solvent and a biological reactant. Because of that, even a slight change in the amount of water inside a human being, 2 percent, say, can cause pale, clammy skin, nausea, confusion, loss of physical strength, elevated heart rate, sleepiness, and loss of appetite. Similar changes in most other components of human beings cause much less dramatic effects, if they have any effects at all.

A slight change in our water changes *who* we are—how we behave, whether we will produce or care for children, whether we will sob, pray, imagine, or perspire. Still we say, water is not a forgotten room inside the house of self—our selves or the selves of any of those who came before us. Water is just passing through us. Water is only the darkness between the stars, never the stars themselves.

———

The first miracle at Lourdes occurred on March 1, 1858. Catherine Latapie—who lived in Loubajac near Lourdes—had recently fallen from a tree. Catherine was thirty-nine and pregnant. (There is no record of what she was doing in the tree.) In the fall, she had damaged her brachial plexus—the nerve plexus that supplies nerves to the chest, shoulder, and arm—and she was left with an ulnar paralysis. As a result, her right hand had resolved into a fist and she could not use two of her fingers at all. One morning at three, moved by a sudden impulse (a dream, perhaps), Catherine with her two children set out for Lourdes. When Catherine arrived at dawn, Bernadette was already there praying. Catherine bathed her hand in the water, and her hand healed.

She returned home, joyful, and that day gave birth to her third child, a boy, who years later was ordained a priest.

———

Water is the one substance on the face of this Earth that naturally exists in all three phases—solid, liquid, and vapor. Water is the only naturally occurring liquid on this planet that expands as it freezes, gains 9 percent of its volume as it turns from liquid to solid. That may seem trivial. But because of that expansion, frozen water floats on top of liquid water—floats upon and insulates liquid water. If water didn't do that, lakes and much of the oceans would freeze from the bottom up—freeze solid. That alone, particularly during the ages of ice, would have destroyed most, if not all, of the living things on this planet.

That expansion below 4° centigrade is also the reason pipes burst as they freeze, bottles of beer explode in the freezer, and the foundations of our homes split. Below 4° centigrade, water is stronger than steel and stone.

———

It is cool, fall along the river. 1986. My wife, Gina, and I are camped for a few days near the Green River in Utah. We have our dog, Nell, with us. By the second morning, I am tired of the plastic Porta-pot we brought with us. So I choose instead to take my morning toilette among the bushes near the river. Finished, pleased with myself and the glorious morning, I set fire to the paper I used. The banks of the river are choked with dried reeds and grasses, and almost as soon as I touch match to paper, the reeds and the grasses around me explode. I scream to Gina to grab the water cans, and for a minute or two we try to carry water from the river to douse the fire. But from the moment it first flickered, it was obvious the fire would win. As the flames begin to turn toward our dog and our truck, we make a run for it. I grab Nell and our still-pitched tent and throw the dog up front, the tent in the bed of the truck. Gina grabs the rest of the gear and joins me and Nell in the cab. I drop the transfer case into low range and grind the first few feet through river sand, just ahead of the fire. Then I switch to higher gears, and we leave the flames behind.

Once we reach higher ground, we stop to catch our breaths and to try to clear the smoke from our lungs. From where we stand, it looks to me as though the river itself is on fire. I am stunned and shamed by that sight. For a million years or more that river lay there grinding rock. Perhaps it has burned before, but I think today is the first time it has ever burned simply because of human stupidity.

Gina says we should leave, find a place with a phone so we can report the fire. It seems a terrible and cowardly thing to leave the river burning here, but what else can we do?

———

On March 25, 1858, the woman-in-the-cloud appeared to Bernadette for the sixteenth time. She was already in the grotto as

Bernadette neared, and a crowd had gathered before the woman. It was the feast of the Annunciation (a celebration of the day the Holy Virgin was told that she was to be the mother of Christ). And for the first time, the woman spoke of who she was. At Bernadette's request, the beautiful woman told those gathered there that day: "I am the Immaculate Conception."

———

There are at least fourteen different forms of water ice. The form that water takes as it freezes depends on the temperature and pressure when the water solidifies. High pressures will generate forms of ice that are solid to near 100°C (the temperature at which water normally boils). You can make this weird ice by forcing water to freeze at temperatures and pressures never found outside of a laboratory, at least outside a laboratory on the surface of this planet.

Beneath the sea floor there is water ice full of methane. If you touch a match to that water, the ice will burn. Some people believe that one day that water will fire the dynamos of civilization.

At a pressure of one atmosphere, the ideal pressure at the surface of the sea, there is only one form of ice—the one we all know—the one that spreads across the lake in winter, the one that doesn't burn, the one that chills our lemonade, the one that finished off the Donner party.

That form of ice is miraculous as well. When water freezes it purifies itself—squeezes from itself nearly every impurity. As it freezes, seawater forces out globs of brine; freshwater squeezes out sulfuric acid, methyl mercury, and salt. And the solid that remains is 99 percent pure water. That's what Eskimos drink—frozen seawater, 99 percent pure water. Regular seawater would kill them. But ice, well, ice is nearly immaculate.

———

It was because of those five words—"I am the Immaculate Conception"—and only because of those five words, that the Roman Catholic Church eventually chose to believe Bernadette and to declare the apparitions at Lourdes miraculous and divine. Only four months before, the Church had officially coined the phrase "Immaculate Conception." It was too soon for Bernadette, a poor country girl, to have known the words of Rome. The five words were finally spoken by the woman-in-the-cloud during her sixteenth apparition. Until that moment, Curé Peyramale, Bernadette's pastor, had denounced Bernadette as a fraud. Those five words nearly changed Peyramale's mind. But he was a priest, so he asked for more.

"Tell this woman," he said, "to make the rosebush that stands beside her grotto bloom. Now, in the middle of winter." He said that to Bernadette because he was a priest and she was a peasant. He said that because he knew that only the warm days of spring could make roses bloom. Warm days and water.

That year the roses bloomed in winter.

———

It is warm and humid, summer again. Gina and I are in Rome for a belated honeymoon. We are very much in love. After we share a dry martini with a friend, he suggests a visit to Villa d'Este in Tivoli. We eagerly accept.

The villa was built in 1500 by order of Hyppolitus d'Este, the governor of the city. It was first a Benedictine monastery, a house of God. Later it was a house of water. The architects of the Renaissance moved the river itself here. And once they had it, those architects sent the river's water through a thousand pipes to light a thousand fountains that turn the air here into wet music and people into giggling lovers.

Perhaps it is only because we are so in love. Perhaps it is only because we are so wet and so in love, but as we walk these lighted, damp paths beneath the Roman sky, I think to myself that I have never seen such a beautiful woman or such beautiful water, such beautiful, old water.

Water rising from green brass spigots set centuries ago into the lawns and walks of Tivoli. Water that was once Napoleon's, was once Cleopatra's, was once Australopithecus's, was once God's. Water that *was* once Napoleon, *was* once God. Only a tiny fraction of the water on this planet was created here or is destroyed here. Most of the water bubbling here has been just as it is now for billions of years, perhaps tens of billions of years.

Rivers of water nearly as old as the darkness itself and reaching all the way to the stars.

——

When water vaporizes it also purifies itself. It leaves behind most all that has contaminated it. Pure water fallen as rain or snow gathers the salts of this Earth as it flows to sea, but then leaves them there as the water evaporates. Then, as the air pulls the water from the sea, water changes to steam and back into storm clouds. Salt water under clouds of pure water. Because of those salts, drinking seawater will kill a person. The salt water will draw more water from us than it gives. Then, surrounded by water, we dehydrate. If we dehydrate throughly enough, we die.

Air will do that to us, too, draw water from us as pure vapor until what remains is no longer us. More slowly, certainly, but just as surely.

The Inuit have some twenty or thirty words for snow. Biologists have ten thousand different names for water—names like bear and beetle, like amphioxus and paramecium, like orchid and fern, like angiosperm and bristlecone, like bloodwort and buttercup.

———

The following is excerpted from one of the most complete descriptions of a miracle at Lourdes.

Jean-Pierre Bély of Poitiers, France, was born August 24, 1936, later married, and, with his wife, had two children. In 1972, Mr. Bély, then an anesthesia and intensive-care nurse, was diagnosed with multiple sclerosis. The disease progressively worsened until 1984, when Mr. Bély was declared "100 percent invalid." In October of 1987, he made his first pilgrimage to Lourdes, France.

During a pilgrimage to the Sanctuary of Lourdes, Mr. Bély regained completely his normal functions, in a sudden, unexpected and unforeseen way on Friday 9th October 1987.

After having celebrated the Sacrament of Reconciliation on the 8th of October in his sickroom, he received the Sacrament of the Sick the next day during the Mass in the Rosary Square during the French Rosary Pilgrimage. Mr. Bély then felt himself overcome with a powerful sense of interior liberation and peace that he never before experienced.

Then, that same Friday, at midday, when he was lying down in the sickroom, he experienced a feeling of cold which grew stronger, almost painful, which gave way to a feeling of warmth, also more intense and overwhelming. He found himself sitting on the side of his bed, and was surprised to be able to move his arms, to feel contact against his skin. During the night that followed, although in a deep sleep, Mr. Bély woke up very suddenly and had the surprise of being able to walk for the first time since 1984. The first steps were hesitant, but it quickly became normal. So as not to appear different from his "companions in sickness," Mr. Bély wished to leave Lourdes in a wheelchair as though he was still an invalid. Arriving at the railway station, he decided, finally, to enter the train alone and to remain seated during the return journey to Angoulême.

The report continues for two more pages, documenting Mr. Bély's cure. How, after he smuggled the miracle out of Lourdes, he carried it to his own home, where he could, in solitude, revel in the joy of having regained the function of his limbs and his loss of pain.

No mention is ever made of the water—the water Jean-Pierre brought with him, the water he found there at Lourdes, or the water he left with. Of course, no mention is ever made of the Immaculate Conception, either. But in fact, at Lourdes the Immaculate Conception was said to have presided directly over, at most, only five of the sixty-five miracles officially recognized by the Catholic Church. Water was present at each.

———

During the summer of 1969, I worked on one of seven teams of men that strung pipe for Halfhill Sprinklers. Each team had a master pipefitter, an apprentice pipefitter or two, a couple of laborers, and a red Ford pickup truck full of galvanized pipe. I was a laborer for team number three. The other laborer on that team was a Mexican-American man of indeterminate age, who called himself Hector. Hector dug trenches as if it actually mattered how trenches got dug, whistled at the women who walked past—all of them—and smoked Lucky Strike cigarettes after everyone else had switched to filter tips.

We talked while we shoveled, Hector and I. He told me stories of the women he had known—the places they found him, the places they left him. He told me about highways through the darkest parts of Texas and the stars that boiled overhead. Once he spoke to me of his son and he told me a story so sad that that night I forced myself to forget it. In return, I told him nearly nothing, because nearly nothing was all I had to tell. I was just out of college, dumb as a rock, and uneasy speaking with men. That seemed okay by Hector.

His talking helped. It helped to turn our minds from the long trenches we opened and shorten the hot hours strung out between morning and evening. It helped to dimple the still pool of boredom that often filled up those trenches.

It was a real job, this one—digging ditches and laying pipe. I had to show up on time and do what I was told to do—punch a timecard, open trenches. A real job. And like every other real job I ever had, I hated it. I hated the regularity of it, the routine. I hated knowing that every morning as I walked into the shop the foreman, Jack, was going to ask me how they were hanging. And I hated how hard the work was.

But unlike the dozen or so other real jobs I'd had, I liked what we did on this job. In spite of how much I hated doing it, I liked what happened when we were done. When all the ditches were filled and the valves were opened on the pipes we'd strung, water moved. It moved just the way we'd intended it to move. And then the water rose into the dry Utah air and made a hundred little rainbows above the Utah dirt. I liked that—moving water from place to place. I liked Hector, too. I liked the way he'd look at the leaking elbows and unions we had strung out and say to me, "Forget about it, boy. It'll water seal. Take care of itself. Forget about it." Then he'd spit on the leaking pipe and we'd bury it. I liked moving water. And I liked Hector.

And then one July day, while I was working in the shop with Hector, wasting words, I got a call about a job in a laboratory at the University of Utah—a job that would ultimately lead to my doctoral degree, and make me, one day, into a biologist. After I hung up the phone, I told the old man about it, the job offer. He wrenched an el onto a spinning piece of galvanized pipe, then spit onto the oily concrete at our feet.

"You gonna take it?" he asked, speaking around his cigarette.

"I don't know," I replied. "The money's better here."

For an instant he hesitated, as though he were about to say something complex. Then he just picked up his pipe wrench and a handful of els and turned to his work. Light the color of lint fell through the window behind him, rain rattled the corrugated tin roof overhead. As he worked, a scar that ran from his left eye to the bottom of his jaw twitched. Finally he stopped for a moment, and turned to me.

"You know something?" he said to me. "You may be one of the sorriest sons-a-bitches I know." Then he went back to wrenching elbows onto eight-foot lengths of half-inch pipe.

I walked back to the phone.

———

On June 2, 1925, Bernadette Soubirous was beatified by the Roman Catholic Church. Two weeks later, a statue of Bernadette (being borne by angels towards the Immaculate Conception) was unveiled in the Basilica of St. Peter in Rome. Eight years later, Pope Pius XI declared Bernadette Soubirous a saint of the Holy Roman Catholic Church and declared that her name be inscribed into the Litany of the Saints. Then Pius himself and everyone gathered at St. Peter's spontaneously sang the "Ave Maria." On both occasions, the beatification and the sanctification, everyone present was blessed with water—the water they came with, the water they found there, and the water they left with—though descriptions of neither ceremony mention it.

The Flame Within

Something about Mary Reeser made Pansy Carpenter's skin crawl. Something about Mary's eyes—deep black like burnt almonds—and Mary's hands—so warm Pansy could barely bring herself to hold them. And that something, whatever it was, made Pansy withhold a sizable piece of her friendship from Mary for fear of Mary herself and what Mary might do to her. At times, this caused Pansy considerable guilt, but she couldn't bring herself to do otherwise.

This was in spite of the fact, which Pansy was well aware of, that all of Mary Reeser's life, most people had spoken of Mary as nothing more than ordinary. Those who felt that "ordinary" was a little harsh said that she was "reliable." And a few others, who were generally less concerned about her feelings, called the woman "unremarkable." But what all of them meant was that Mary Reeser was predictable as tree sap. Mary reveled in her routines.

Most afternoons she took her straw hat and went for a walk from her home at 1200 Cherry Street Northwest in St. Petersburg, Florida. Nearly every weekend, she visited with her son, Dr. Richard Reeser, and her grandchildren. Every Wednesday morning she did laundry and bought groceries. Each Friday afternoon she tatted lace, and every Sunday evening she read her Bible, usually Acts.

Sixty-seven and widowed, Mary Reeser was a lantern over a

sea of change—the kind of woman you could count on to be where you expected her to be when you expected her to be there. And Pansy, her landlady, knew Mary as well as anyone did. Pansy knew Mary's routines, knew where to find her when she needed her, knew just when to expect the monthly rent check, and nearly knew which dress Mary would choose to wear before Mary herself knew. Mary was dependable. Nevertheless, there was something about her that made Pansy more than a little uneasy.

Sunday night, the first of July 1951, was no exception. To the east, the downtown St. Petersburg skyline periodically snapped into sharp relief beneath summer heat lightning and the air smelled of rain. Pansy sat on the patio behind the apartment building and watched the storm gather itself. She shivered slightly as thunder stumbled across the intervening spit of land. *It will be raining soon,* Pansy thought. *I should get inside.*

On her way to her apartment and her bed, Pansy stopped at the door to apartment E to see how Mrs. Reeser was doing. Pansy knew Mary was upset because friends of hers had not yet found her an apartment in Columbia, Pennsylvania, and Mary was anxious to return to Pennsylvania. The door was open, and Pansy took that as an invitation to enter. Mary was in her favorite chair, her pajamas already on, smoking the evening's final cigarette. It pleased Pansy to find her there.

"Looks like rain again tonight, don't you think, Mrs. Reeser?"

"Yes it does, Mrs. Carpenter. I surely hope it doesn't ruin the petunias."

"Oh, I don't imagine it will rain that hard, dear. I wouldn't worry. Did you have a nice time with your grandchildren today?"

"I did, Mrs. Carpenter, I surely did."

"I'm glad to hear it, dear. Try not to worry too much about things in Pennsylvania. I'm sure it will all work out."

"I'm sure it will, dear."

"Well, I'm off to bed. I hope you sleep well."

"I will, Mrs. Carpenter, I'm certain. I took two of my sleeping pills and those always do the job. I hope you sleep well, too."

The two women smiled at one another and then to themselves. Pansy gave Mary's hand a squeeze and started down the carpeted hall. *Something about that old woman,* Pansy thought to herself as she rubbed her hand where she had touched Mary Reeser, *something queer.* And once again Pansy felt a pang of guilt. But as great a thing as guilt is, it didn't chase the warm touch of Mary Reeser from Pansy Carpenter's hand or the uneasiness from her heart.

The last of the smoke from Mary's cigarette finally fell behind as Pansy closed the door to her own apartment. Inside her bedroom, Pansy pulled off her dress and brassiere. Tossed her bra into the hamper and hung her dress in the closet. She stood for a moment in nothing but her panties, looking at herself in the mirror on the closet door. She sighed heavily, then pulled on her pajamas. Mechanically, Pansy climbed into bed and pulled the pink coverlet up to her chin.

Guilt took one last charge at her, but Pansy's pictures of Mary Reeser still seemed a little off-center, a little out of focus. Tomorrow Pansy would suggest a trip to the botanical gardens— Mary always seemed to enjoy that. That thought eased Pansy's guilt some. She sighed to herself and felt a little better. But as sleep finally claimed her, the word *peculiar* opened and closed once more like a brass hinge in Pansy's mind.

Like most of us, Pansy Carpenter took many things for granted that night—the water in her taps and toilets, her son-in-law's love for her daughter, the moon, the pulse at her wrist, the fact that one day she would remarry, and the trip next Tuesday with Mary Reeser to the botanical gardens. All of this, Pansy imagined, as most of us do, would seem as probable tomorrow as it did today. With that in mind, sleep came easily.

But as Pansy began snoring deeply on that first evening of

July, 1951, Mary Reeser was already dead. And neither Pansy nor anyone else would ever understand how or why.

———

Distinguishing between a dead thing and a living thing is not as simple as most people imagine. Things that appear obviously alive—like the ice crystals growing across a winter windowpane or the particles dancing with Brownian fervor in a drop of magnified pond water—are dead things. And things that appear obviously dead—like the black scars on mountain granite or the red stains beneath glacier ice—are live things.

In spite of what most of us might wish to believe, it is no great leap from the isle of the living to the abyss of the dead. There is no indelible mark left by a living thing, no unmistakable hue to the dead—at least none that rises in human eyes. The living and the dead are as near to one another as our bodies are to their images in our mirrors.

But they are not the same things. The living are not the dead. Something irrevocably isolates the two. That is obvious, to all of us, even our children. But the thing itself, the thing that divides living animals from dead animals, is a very small thing, such a small thing that it was not even noticed by men and women until a hundred or so years ago, when microscopes were invented. And just what this thing actually did wasn't understood until fifty years ago—unknown billions of years after it first slipped inside of some animal's cells.

It is the reason for the warmth of a human touch, the ember inside a human kiss. It is the heat of a quickly drawn breath, the blue pulse inside the heart's red sheath. It is the reason we sleep with dogs, build houses, weave cloth, and hold one another. It is the light of life itself. When that fire first flickered isn't known, but it was mothers, ours and others', who kept the lamp ever since and kindled each new flame.

––––––

Each human begins life as a single cell, a zygote. That zygote comes to be when a father's sperm, racing up a fallopian tube, penetrates the zona pellucida of a mother's egg. To that zygote the sperm and the egg each gave twenty-three chromosomes. A human's DNA. That DNA is the blueprint for what he or she will become—half from the father and half from the mother. Equal partners, they were, in drawing up the plan for our lives.

But the lamp of life—the lamp that lit the plans for the architect and the builder, the lamp that warmed the carpenters and the plumbers, the cabinetmakers, the electricians, the landscape architects, and the new owner—that lamp was passed to us by our mothers. Only our mothers. And it has been so passed for more years than any human can possibly imagine.

Sperm are little packets of DNA to which tails have been attached, tails that drive sperm towards their goals. A sperm's only goal is a woman's egg. And a sperm's only role in life is to fertilize that egg. Sperm are very good at what they do. They are not much good at anything else.

Specialization always comes at a price. Sperm have a head full of DNA, an enzyme or two, and a tail. A nucleus and a tail. That's it. Then at the critical moment—the moment of fertilization—the sperm loses its head to the nucleus of the egg and leaves its tail snapping uselessly at the point of collision. Not much of a life. But it serves its purpose, and on a good day in the warmth and darkness of a fallopian tube, a sperm delivers twenty-three chromosomes to an egg.

The egg, on the other hand, comes to this union with a little less haste and everything else needed for a long and happy life. That is, the egg comes not only with a nucleus and twenty-three chromosomes, but it also comes with its cytosol—all the stuff that surrounds the nucleus. In that stuff, the clear liquid of the egg, there are a thousand little flames burning with the fire of the sun. The lamp of life, a mother's gift.

———

The next morning at five A.M., Pansy was awakened by the smell of smoke. At first she was frightened and confused. But then she remembered that the pool pump in the garage was failing, and twice before that pump had smoked like this. Groggily she found her way to the garage and turned off the pump. Just as groggily, Pansy stumbled back to bed and, after making herself a note to call the repairman in the morning, quickly fell asleep. But at eight A.M., she was awakened again, this time by the doorbell. *What is it now?* she thought as she reached for her watch and her house slippers. *Who could be at the door at this hour of the morning?* Pansy pulled on her housecoat and made her way downstairs. She reached the end of the hall and opened the front door of the apartment building.

"Yes, what is it?" she asked, a little curtly, of the uniformed young man standing at her front step.

"A telegram for Mrs. Mary Reeser, ma'am."

"Very well," she said as she took the yellow envelope from his hand and began to close the door.

"Excuse me, ma'am," the young man exclaimed as he reached to hold the door open, "I need someone to sign for that telegram."

"Oh," she said, as she released the telegram from her breast and moved back toward the door. "Very well, then. Do you have a pen?"

Pansy took the pen he offered and quickly scrawled her name across the bottom of the young man's clipboard.

As she wrote, she wondered who might have sent the telegram. *But whoever sent it, the telegram certainly isn't good news. Telegrams never are.*

Pansy carried the envelope down the hall toward Mrs. Reeser's room. *She should be up by now,* Pansy thought. *Mrs. Reeser is an early riser.*

She rang the bell on Mrs. Reeser's apartment door. Nothing. She waited a full minute—no answer. She rang again. A minute more. Still no answer. Once more she rang the bell. She leaned her ear against the wooden door and held her breath. No sound came from within. Pansy knocked as loudly as she could.

"Mrs. Reeser? Are you in there?" Still there was no response. Finally—and mostly without thinking because Pansy just didn't do that sort of thing—she reached for the door handle.

Pansy grabbed hold of the brass knob, and then instantly snatched her hand away. The doorknob was hot. The doorknob was very hot!

———

What our mothers brought many of to the union of egg and sperm and what our fathers brought very few of are mitochondria—little peanut-shaped organelles about a thousandth of a millimeter long. Inside of each mitochondrion is the living flame—the flame that separates living humans from dead humans.

One thousandth of a millimeter is about a hundred times smaller than the diameter of a human hair. Invisible, even with a reasonably powerful microscope, these organelles are nearly everywhere inside every one of our cells. Mitochondria are the keepers of the flame. And the heat of their fire has driven every human act, ignited every human thought, and inflamed every human love since anything remotely human shuffled its feet across planet Earth. Mitochondria make dead things into living things.

For no apparent reason, after fusion of egg and sperm, the

cytoplasm of the zygote quickly destroys the few mitochondria that arrived with the sperm. So the only mitochondria that survive are those that came to us from our mothers. And of course, our mothers' mitochondria came from their mothers, and so on. But beyond that, the trail has grown cold.

Our mitochondria, the ones floating around inside our cells, look more like bacteria than they look like anything else. Mitochondria have their own DNA, their own genes. Mitochondria divide more or less whenever they please. And when they please is in no way related to the division of our cells. The DNA of mitochondria is rolled up in a circle, like the DNA of bacteria. Mitochondrial ribosomes, the little machines that make proteins, look like bacterial ribosomes. Mitochondria are surrounded by two sets of lipid membranes, like bacteria. And none of them divides like our cells do. They divide like bacteria do.

So it is most likely that the things we call *our* mitochondria—the things that have powered every single human thought, the things that have fanned the flames of every human love—were once free-living bacteria. Oddly enough, the same bacteria that later came to cause typhus. Scientists can tell that from the DNA inside of mitochondria.

It is impossible to say how they got there—inside of us, that is. But if mitochondria began life as free-roaming bacteria some two or three billion years ago, they could have gotten into the cells—that would one day be our cells—when the cells tried to eat the bacteria or when the bacteria tried to infect the cells. In truth, it makes little difference. But once those bacteria had crawled inside our cells, an odd thing happened. The bacteria discovered that our cells provided an unbelievably rich source of nutrients. Our cells, on the other hand, discovered that the bacteria provided unimaginable bursts of energy. A contract was signed, a match was struck, and a tiny flame flickered inside of cells that would one day be human.

———

Pansy cried out as she jerked her hand from the doorknob and she jumped back into the hall. Two painters working outside, Albert Delnet and L. P. Clemens, heard her cries and ran in to see if they could help. Pansy stood with her hand to her mouth and just pointed at the doorknob. The first of the painters gingerly reached for the knob. He drew his hand back quickly.

"The knob's hot," he muttered.

"You've got to open that door," Pansy said. "There's an old woman in there who needs help."

"Stand back," the second painter said, and he took a run at the door. His shoulder bowed the door inward sharply, but the door stood.

"Once more," he said and took a second run.

This time, the hinges pulled from the wall with a sound like a shot, and the door fell inward. The room exhaled hot air.

There was smoke, but less than the men had expected. They both raced through the apartment looking for Mary Reeser. But after a few minutes, they could find no sign of her. The men, coughing, stepped back into the hall.

"She's not in there, ma'am," the first man croaked to Pansy. "Her bed's empty."

"How can that be?" Pansy asked.

"I don't know, ma'am, but I think we should call the fire department. There's likely more fire than what we see."

"Yeah," said the other man, "sometimes it burns way up inside the walls."

"Oh dear," Pansy said and sat heavily on the hall settee.

———

The single thing that most distinguishes living things from dead things is the ability of living things to convert energy from one form to another. Plants make sunlight into sugar; herbivores make plants and their sugars into the mechanical energy of motion; and omnivores and carnivores make herbivores into fear and flight, eyesight and thought, heartbeats and hope. Every living thing makes its living by drawing energy from its surroundings and converting that energy into useful things like protein, fat, nucleic acids, and hope. That's life.

Plants store solar energy in two places—sugars and fats. The energy extracted from sunlight by chloroplasts (also likely to have been free-living bacteria once) is used to make sugar molecules. Some of that energy is then transferred from sugar molecules to triglycerides, for long-term storage in fat molecules. Animals that cannot extract energy directly from sunlight live off the energy earnings and storage of others. That is, men and women rely almost exclusively on fats and sugars from the animals and plants they eat for the energy needed to do what humans do. And inside human beings, it is the mitochondria that strip the energy from the fats and sugars we eat and make that energy into things human.

Inside our cells, enzymes convert the sugars we eat into a substance called pyruvic acid. Pyruvic acid, in turn, becomes acetyl coenzyme A. Fats that we eat are also turned into acetyl coenzyme A. This is important, because acetyl coenzyme A is mitochondria food. What mitochondria do better than anything else in the world is to convert acetyl coenzyme A into oxaloacetate, which, with the addition of a little sugar or fat, will shortly be made once more into acetyl coenzyme A. This set of biochemical events is called the Krebs cycle, since it was first described by Hans Krebs in 1937. Converting acetyl coenzyme A into oxaloacetate and oxaloacetate into acetyl coenzyme A, and so on, in spite of its symmetrical purity, seems a little senseless.

But the circular part—acetyl coenzyme A to oxaloacetate to acetyl coenzyme A and so on—is only what is happening on the surface of things. While acetyl coenzyme A is being turned into oxaloacetate, carbon atoms—the stuff of coal and Number 2 pencils—are being trimmed off, and mitochondria oxidize, or burn, those carbon atoms into carbon dioxide, or CO_2. That burn produces a little extra energy that can be used for other things inside of cells. But that energy, the heat given off during the conversion of acetyl coenzyme A to oxaloacetate, is way too little to fuel the processes of life for anything as large as a human being.

For human lovemaking or bicycling or writing, more is needed and mitochondria provide that more by burning carbon atoms to drive the reduction of a molecule called nicotinamide adenine dinucleotide, or NAD. NAD is a coenzymatic form of the B vitamin niacin. NAD helps strip electrons off sugars and fats, the high-energy molecules we eat, and it turns those electrons into energy we can use. In effect, inside of mitochondria, NAD takes fats and sugars and generates hydrogen. Hydrogen, the sun's own fuel. Hydrogen, the same stuff the zeppelin *Hindenburg* was full of when it struck its metal mooring, discharged its huge load of static electricity, and went up like napalm.

———

When the firemen arrived, they quickly extinguished the only fire that they could find—a small flame burning inside a wooden beam that separated Mary Reeser's kitchen from her living room. Then they tore away a couple of partitions to check for further fire. None was found. As the smoke cleared from the apartment, Fire Chief Griffith noticed an odd-looking pile of charred rubble in the middle of the living room. The spot struck the fire chief as odd because as far as he could tell, it was the only spot in the entire room, perhaps the entire apartment, that had obviously burned.

He walked over and began to poke with his toe at what lay in the middle of that four-foot diameter of charred flooring. At first, the only things Chief Griffith recognized were scorched steel coils that looked like chair springs. Then the chief knelt and began to sort through the rest of the rubble with his ungloved right hand. He lifted one scorched item, then another. He recognized nothing. He reached for another large piece and lifted it, brought it to his nose. Parts of it were soft and smelled a little like burnt steak.

Chief Griffith then held the piece in his hand at arm's length. It took shape. It was most of a liver stuck to a piece of spine. The chief bit his lower lip and dropped what he held. On the floor next to where it fell, there was the calf and foot of a human leg. A black slipper still covered the foot. Next to that, a bit of skull. Chief Griffith drew back and gagged hard.

———

When the *Hindenburg* went up like, well, like the *Hindenburg,* the fire killed thirty people, torched over seven million cubic feet of hydrogen, and destroyed the dirigible itself, the mooring pad, and acres of land. Clearly, mitochondria don't burn hydrogen in the amounts or in the manner of the *Hindenburg*. If they did, humans (if there were any humans at all) would be regularly going off like bottle rockets. Instead, mitochondria have evolved an elegant series of steps for burning hydrogen—a waltz that cools the burn and prevents most of what finished off the *Hindenburg.*

The first thing that mitochondria do to avoid the zeppelin's fate is to strip the hydrogen off NAD and split it into protons and electrons. The protons get stored much as protons are stored in batteries, and the electrons get shuffled through about fifteen different molecules called electron acceptors. When hydrogen is combined directly with oxygen, a huge amount of energy is released all at once, more energy than any human cell could possibly handle. What the electron acceptors inside of mitochondria do is to release that energy a little at a time. Like slowly rolling a pumpkin down a set of stairs, rather than simply dropping it from the third floor. Each of the fifteen electron acceptors has a slightly greater affinity for the electrons that NAD originally stripped from the sugar or the fat that we ate. Because of that, each acceptor extracts just a bit of the energy that would be released explosively if the hydrogen was just burned to water and then passes the electron on to the next acceptor in the line. Near the end, the potential energy of the stored protons is used to make adenosine triphosphate—the energy source for everything humans do. And in the final step the hydrogen's protons and electrons are recombined with one another and with oxygen to make water, H_2O—water from fire.

From one-half cup of sugar, mitochondria extract enough energy to boil three quarts of water. From one-half cup of fat, mitochondria extract enough energy to boil three gallons of water. And all of this is ultimately driven by the oxygen we breathe and its affinity for the electrons of hydrogen—the same affinity that torched the *Hindenburg*.

———

The police were called. Fire Chief Griffith was certain they should be. Pinellas County Coroner Edward Silk arrived with the police. Mr. Silk poked through what was apparently the remains of Mary Reeser. Mrs. Carpenter told the chief that she had seen Mary wearing those very slippers as Mary finished her cigarette last night. The coroner was baffled. This was unlike anything he had seen in his twenty-two years on the job. Lying inside the circle of charred flooring were the coil springs (all that remained of Mary's overstuffed reading chair), the pieces of Mary Reeser's leg, spine, and liver, and what appeared to be an intact human skull, not a piece of skull as Chief Griffith had originally thought, but a whole skull reduced to about the size of a baseball.

Coroner Silk couldn't even begin to fit the pieces together. He continued to sift through the remains but found nothing else. He looked around the room, he spoke with Pansy, he looked at the still-locked apartment windows, he looked at the locked door. There was no evidence that anyone but Mary had been in this room last night. No evidence that anyone had come in through the windows or the door.

Finally, the coroner wrote "smoking accident" beneath the words "Cause of Death" on the form stuck to his clipboard. He wasn't happy with that, but everything else that came to his mind could have cost him his job. Coroner Silk then authorized the removal of Mary Reeser's remains to the hospital morgue, and he left.

In spite of his efforts to defuse the whole situation and save himself from ridicule, Edward Silk's final report would eventually become what is today known as the mysterious case of Mary Reeser. This case is still considered by many to be the best-documented account of spontaneous human combustion—including the fire-scene photos of Mary Reeser's single leg and slippered foot. I've taken a few liberties here—added what I imagined were the thoughts and words of those involved, described a little more about people's behavior than is actually known—but the facts of the event, as described by the people who investigated Mary Reeser's death—people like Chief Griffith—appear here just as they first appeared in the written reports of this bizarre accident.

The living room was marked in an odd way. None of the furniture, other than the chair and the end table next to it, was badly burned. But the ceiling, draperies, and walls—from a point four feet above the floor—were covered with an oily yellow soot.

Below four feet, the walls looked untouched. And when the investigating lieutenant cleared away the last of the ashes, the carpeting below where Mary had died was only superficially charred. On the adjacent wall, ten feet away, a mirror had cracked; and on the opposite wall, twelve feet or more from the body, two pink candles had puddled in their holders, but their wicks were intact. Four feet above the floor, the wall outlets had melted, but none of the fuses had blown. And the outlets in the baseboard appeared untouched.

The hospital lab reported that all that remained of Mary Reeser amounted to about ten pounds—ashes, skull, leg, liver, and bone. When asked, Pansy Carpenter guessed that Mary had weighed somewhere near 175 pounds the night she died. Mrs. Reeser was not a frail woman.

An arson team, led by Edward Davies from the National Board of Underwriters, investigated as well, but the team could find nothing to add to what the coroner and the police had found— no evidence of accelerants, no evidence of foul play. In spite of that, everyone agreed—nothing about this fire appeared normal. Apartments and people just don't burn like that.

———

But we do, each of us, burn. We burn with a billion flames. Fires given to us by our mothers, fires that will smolder in each of us for a lifetime. Two things control the rate at which those fires burn: the mitochondria themselves, and in humans, the thyroid gland. The thyroid gland, which is in turn controlled by the hypothalamus of the brain, produces a hormone that acts like a thermostat and turns the mitochondrial flames up and down as needed for human function. It seems likely there are other regulators as well, but none has been identified clearly.

As the fires burn inside our mitochondria, most of the energy released goes into making new protein, fat, and carbohydrate molecules, moving muscles, transmitting nervous impulses, speaking, breathing, walking, making love, being human. But about 30 percent of the energy unleashed by mitochondria is just blown off as heat. Normally that heat simply radiates from the surfaces of our bodies. That's the reason we glow like Saint Elmo's own when we are viewed through infrared glasses. But inside a sleeping bag or a wool sweater, inside a pair of gloves or a down jacket, beside a dog or inside a house, the heat "wasted" by mitochondria warms us and allows us to live in places humans could not otherwise survive.

Undoubtedly the St. Petersburg Police Department wasn't short on opinions. Surely there were some who called to offer, off the record of course, explanations for Mary Reeser's death. Space aliens. People had seen it before. Or maybe the fire of hell itself destroyed Mary Reeser that night. From the records, though, it appears that many who came forward agreed that there was a simple explanation. Mary Reeser, they said, had spontaneously combusted, gone off like old oil- or gasoline-soaked rags left too long in the garage. Spontaneously ignited, they said, when her body's chemistry crossed some unknown boundary. And alcohol, some of them argued, added to it. Alcohol and the ember of her cigarette. Sometimes that's all it takes to cross the line between the normal slow burn of life and the explosiveness of a torch. A lethal combination, they said, and it happens all the time. But there was no evidence that Mary Reeser was drinking that night. No evidence, in fact, that she ever drank. And the forensic pathologist involved, Dr. Wilton Krogman, who spoke for all of the sane and scientific, said that while he had never seen anything quite like it, there was not one shred of evidence to suggest that Mary Reeser or any other human being had ever spontaneously torched. But the evidence was confusing, so confusing, in fact, that J. R. Reichert, the chief of police in St. Petersburg, even asked the FBI to investigate. "It is not generally realized the extent to which a human body can burn . . ." the FBI wrote in its final report. "It was formerly believed that such cases arose from spontaneous human combustion or the burning was sometimes attributed to preternatural causes. There is, however, absolutely no evidence from any of the cases on record to show that burning of this nature occurs." Nevertheless, even some of the police came to suspect spontaneous combustion, but of course, they never spoke of it to the newspaper reporters. After all, who could prove such a thing?

The fire that burns inside mitochondria is spontaneous. There is proof of that. Chemists have a way of determining if a reaction is spontaneous or not. I mentioned earlier the Gibbs Free Energy. The Gibbs Free Energy provides a way of measuring whether a chemical reaction produces or requires energy. If the Gibbs Free Energy of a reaction is negative, then the reaction produces energy and is a spontaneous reaction. If the Gibbs Free Energy of a reaction is positive, then the reaction will not proceed spontaneously. These reactions must have energy pumped into them before they will proceed. Lawn mowers and chain saws are driven by spontaneous reactions—the energy-producing reaction of hydrocarbons (gasoline) with oxygen. Refrigerators and vacuum cleaners are not, at least not as they function inside of our homes. Refrigerators remove warm air from cold spaces. That does not happen spontaneously and can only be made to happen by the input of significant amounts of energy (110 volts AC).

The Gibbs Free Energy of the reactions that occur inside mitochondria are negative—their reactions, their flames are spontaneous. Mitochondria's business is spontaneous human combustion.

Lying in her bed a week later, Pansy Carpenter shivered and couldn't decide what to think. Pansy had known Mary Reeser for over two years. In all that time Pansy had never seen Mary do anything that anyone, by any stretch of the imagination, might call extraordinary. Until the night of July 1, 1951. The idea that Mary had spontaneously burst into flame while smoking a cigarette in her own living room was more than Pansy could possibly believe. But neither could Pansy think of anything else that might explain how Mary died. Certainly Mary had not died, as Edward Silk had suggested, in a smoking accident. *He must think I'm an old fool,* Pansy thought to herself.

Again a picture of the last bits of Mary Reeser scattered among the ashes of her reading chair rose before Pansy's eyes. *How sad, how terribly sad. And what a waste,* she thought.

The heat that July evening was exceptional even for Florida, and Pansy's air-conditioning struggled. Pansy pushed her sheets off and turned on the fan on her bedside table. She could almost feel the flames rising from her own body. And to make matters worse, Pansy felt as though she was catching cold. She pulled the thermometer from its case on the table, held it beneath her tongue for exactly three minutes, and read it. One-hundred-point-one. It was going to be a long night.

But as the air from the fan cooled her slightly, inside Pansy's head there rose a bubble of vindication. And as that bubble rose, it pushed aside every other one of Pansy's thoughts. When at last the bubble burst, there were Mary Reeser's burnt-almond eyes and her dreadfully warm hands. And that was enough. With the glow of righteousness in her old eyes, Pansy Carpenter said to no one in particular, "I was right. All along, I was right. There *was* something queer about that old woman, something dreadfully queer." Then Pansy Carpenter crossed herself as she lay back in her bed and prayed to the Holy Ghost to guide and protect her.

◉

Madness

There are clouds in the painting, of course. Almost any one of us would have included those clouds, thick with electricity and rainwater. And there is the wheat field, smudged out like an empty palm, orange beneath the storm-stricken sun. Surely, many of us would have insisted on the wheat as well. Through the middle of the wheat, a rutted road slices to the horizon and disappears beneath the clouds. Even I, a scientist, would have included the road. A storm like that demands a road. Without the road, there is no hope at all.

But then there are the crows—the one true hint of what had been and what was to come. Crows, fistfuls of them, flung into the swirls beneath the angry wet anvils. All that he had lost, irretrievably lost, he put inside of those crows.

———

Van Gogh died because of an instant, or a lifetime, when the portrait of his life appeared worse than the portrait of his death. Died because his pictures filled up with crows. We call that a behavioral disorder, because we imagine healthy people don't see the crows, healthy people don't choose death over life. And we say that these behavioral disorders are caused by mental illnesses—"mental" illnesses—to distinguish them from real diseases: infections, tumors, broken bones, burst blood vessels, polio. Real diseases, diseases of the *body*.

We do that—razor medicine off at the neck—because of people like René Descartes and Pope Urban VIII and their agreement that the human soul resides in the human mind, and human disease resides in the human body. Sometimes because of that, we believe people with mental diseases are less genuinely "diseased" than people with somatic illnesses. Sometimes we even believe that people with mental diseases and behavioral disorders suffer more from weaknesses of spirit, flaws in their characters, than from genuine diseases. Beneath our collective breath, we say that the crows are all inside their heads, and having said that we imagine that the crows are not real.

———

My uncle Henry had a habit of leaving his fly unzipped, completely unzipped regardless of who might be available to notice. My mother, his sister, hated that. Henry's underpants were usually urine-stained, his shirttails hung out of the opening in his pants, and he had a propensity to yell "Shit" and spit for no apparent reason. Mom hated that, too.

And because Mother imagined that Henry's eccentric behavior was concrete evidence of his total disregard for others, especially his sister, my mother raised up a little hatred for Henry himself. Over the years, that hatred blossomed inside her and bore seed.

I don't think Henry ever noticed how much my mother despised him. At times, I'd watch him look at her with his sea-blue eyes, and I'd see something back there in the hollows. Whatever it was, though, it wasn't shame, or resentment, or anger, or even understanding.

Henry's been dead now for well over thirty years, but my mother still gets angry whenever I mention his name. I don't understand that. Henry, I imagine, tried my mother's patience at times. But I'm certain he never intended to haunt her for three decades after his death. In fact, I don't think Henry intended much of anything, at least not toward the end.

My mother, though, is still tied to Henry, thirty-seven years after he cut his ties with everything and was laid to rest. And if you asked her today why she so despises Henry, my mother would tell you it is because he was dirty, and he was crazy. What she wouldn't tell you (but you might discern if you listened to her for a moment or two) is that she still believes if Henry had cared to he could have stopped being crazy just as easily as he could have stopped being dirty. His dirtiness and his craziness were just Henry's way of getting to her, just his way of making her life more difficult.

INSTRUCTIONAL FILM, SCENE 1: A bucolic panorama somewhere in, say, Nigeria. The sky is first-morning blue, and a breeze tinged with woodsmoke is ruffling the tall grass. There are cattle in the grass. A few of them are grazing, but most are standing or lying down and lazily swatting flies with their tails. At first, the cattle are the only animals we see. But as the camera zooms in, we notice that among the grasses where the cattle are sitting or grazing, there are brown ants and a few land snails. At first, it seems that the ants and snails might risk being eaten by the cattle. But as we watch, we see that both ants and snails always stay low enough in the grass to avoid the foraging ruminants—idyllic, mutualistic, Nigeria. Fade to black.

INSTRUCTIONAL FILM, SCENE 2: The camera's eye reopens a few miles to the north. The sun has fallen beyond the horizon and the breeze that is wrestling with the grass has cooled noticeably. Most of the cattle here are grazing. As the camera closes, we notice something oddly different from scene 1. Below the cattle, many of the ants have worked their way up the shafts of grass and appear to be waiting for something. As we watch in cinematic magnification, a pink tongue, large as a python, wraps itself around several insect-encrusted blades of grass, and brown ants, lots of them, suddenly disappear behind a huge set of cud-scarred teeth. Again, fade to black.

Let's consider, for a moment, what we have just witnessed. In the first scene, the ants are cautious, responsible, sane. In the second scene the ants are none of those things. Because of that, we are tempted, briefly, to say that the ants in scene 2 are mentally ill. But we don't say that, because we don't imagine that ants have minds—at least not like humans do. By definition, ants cannot be "mentally" ill. Whether that is true or not, it serves a purpose, it cuts away some of the gauze that surrounds behavioral diseases in humans. That's useful.

Because, whether we are comfortable with the words or not, the ants in the second scene clearly *are* crazy. They have climbed where they know they shouldn't and remain there with total disregard for the danger. They have lost the ability to care for themselves and seemingly no longer regard life more highly than death. These ants are insane, deranged, imbalanced, nutso. Men and women who behave similarly line the lavatory-colored halls of our country's mental institutions from Passaic to Seattle.

Still, we don't look for histories of child abuse among ants, discuss ant toilet training, or accuse ants of character flaws and laziness. And we don't imagine that ants' problems are all in their heads or call ants crazy. Because, after all, these are ants. So if "mentally ill" isn't accurate, what should we call them?

Surprisingly, the answer to that question has come, not from psychologists or psychiatrists, but from microbiologists, specifically parasitologists. And tedious dissections, not discussions, showed the way.

In the second scene (the one with the self-destructive ants) there was another actor, one we couldn't see, at work inside of every character portrayed. Because it is a very small actor, it is, perhaps, understandable that it escaped our attention. Everyone in the second scene was infected with a microscopic, lance-shaped fluke—a trematode, by the name of *Dicrocoelium dendriticum.*

D. dendriticum is a parasitic flatworm. Parasitism is one of the oldest and most venerated ways of life on this planet. Living things have evolved to parasitize nearly all other living things— plants, animals, microorganisms, even parasites themselves are often parasitized by smaller but equally devious life forms. And the animals and plants that do the parasitizing are as varied as their hosts. All of the things that parasitize animals fall into two groups—protozoans and helminths. Protozoans are single-celled animals like *Plasmodium falciparum,* the protozoan that infects nearly one-third of the world's population and causes malaria. Helminths are worms—round, flat, and tape. *D. dendriticum* is a flatworm, or a fluke, as flatworms are sometimes called.

In spite of their amazing variations, parasites all have one characteristic in common—they cannot reproduce themselves outside of their hosts. All parasites have lost the ability to perform one or more vital functions, usually having to do with collecting or digesting food. As long as parasites are inside an animal's body, their hosts take care of these functions for them. Outside their hosts, parasites often die.

Many parasites have evolved complex life cycles which help (in ways only parasites understand) to meet this uniform need. By complex, I mean many parasites' life cycles involve several different hosts and several different stages of parasite development. Necessarily, this cycle feeds back on itself so the parasite ends up in the same host where it began the cycle, and the whole process starts over.

D. dendriticum is like that. In fact, *D. dendriticum* has one of
the more interesting life cycles of any parasite.

Life begins for this fluke in the bile ducts of grazing cattle.
This is where the adult flukes lay their eggs. Bile is produced in
the liver and is transported, via the bile duct, to the intestine,
where the bile aids in the digestion of dietary fats. In infected
cattle, when the bile moves from the liver to the intestine it takes
D. dendriticum eggs along with it. A short while later, the eggs
find their way—along with the cow's or bull's feces—onto the
grasses underfoot. There snails, with inexplicable tastes, ingest,
along with the cattle's feces, the parasites' eggs. Inside of the
snails, the eggs of *D. dendriticum* hatch, pass through two spo-
rocyst stages of their lives, and then transform into another life
stage, called cercaria, and migrate to the respiratory chamber of
the snails. Inside of their respiratory chambers, snails make slime
balls to aid their movement across the fields. When the slime
balls are secreted onto the snails' feet, so are the cercaria. As the
snails then make their way toward whatever it is that draws
snails, the slime balls are left behind in the grass.

Slime balls are ant food. When ants eat these slime balls, they
also eat the cercaria of *D. dendriticum*. In the ants, *most* of the
cercaria encyst in the walls of the ants' abdomens. But inside of
each ant, one or two of the cercaria migrate to the ant's head
and encyst in the subesophageal ganglion, a part of the ant's
brain. Here the cercaria transform into another life stage, called
metacercaria. Unlike those metacercaria left behind in the ab-
domen, these never become infective. These metacercaria do
something else; they drive their hosts mad.

As evening approaches and the air temperature drops, ants
infected with *D. dendriticum* do not return to the colony along
with their fellow workers. Instead, the infected ants climb to the
tops of the surrounding grasses, clamp their mandibles into the
grass blades, and remain there, immobile, until the morning sun

warms them again. When that happens, the ants resume their normal behavior—at least those that survive do—until the following evening.

Temporary insanity.

Temporary because it lasts only as long as the sun is down. Insane, because the timing of the ants' indiscretions corresponds exactly to the feeding cycles of the grazing cattle who feed most vigorously during the late evenings and early mornings. But here, the grasses are filled up with mad beings that suffer not from poor toilet training or moral and spiritual turpitude, but from an infectious disease.

Parasitic madness. Madness with a past and a purpose.

Each night beneath the African moon, crazy ants climb to their perches atop the grasses of Nigeria and wait for the cracked molars of hungry cattle to end the ants' mad ritual. When the ants' madness is complete and they are finally eaten, *D. dendriticum* completes its complex life cycle, and the arduous trip from cow to ground to cow is closed once more. Inside the cow, digestive juices strip ant from parasite, and while the scene fades to black, life begins again, minus a few crazy ants.

———

In the jungles of South America, there is another dance that couples ants and parasites. This parasite is a mushroom. Beneath tropical canopies, spores of Cordyceps, a mushroom, are whipped about on warm equatorial breezes, and spun between tree trunks and twisting vines until these bits of life land in the spiracles of black ants. Spiracles, holes in the tough exoskeleton of ants, allow ants to breathe. Cordyceps uses the spiracles to get beneath ants' skin. Once inside the ants, the fungus attaches to the soft tissues and begins to raise a family. For a few days everything is fine. But the fungus knows that it will need more than the ants can provide. Soon, deep within the ants, the fungus will achieve sexual maturity, and it will be time to sporulate. Once again, the infected ants develop a sudden acrophilia. Driven by their new-found urgency, the ants leave the relative safety of the earth, climb to the tops of the grass, clamp their mandibles onto the tips of the green shoots, and hang there. The fungus then consumes the ants' brains and sprouts through the ants' emptied skulls. Bathed in sunlight, once again, the fungus sporulates. At the grass tops, where the wind blows freely, the spores are quickly spread, sometimes for miles, but always to other ants. The fungus has lifted itself from the primordial slime, gathered itself upon the wind, and set off, once more, for a new life.

Other varieties of Cordyceps mushrooms parasitize and alter the behavior of caterpillars, mealybugs, and beetles. Fungal madness. Infectious insanity.

———

And then there's the odd story of *Wolbachia*. Most of us, I think, believe that genetics and evolution pretty much predetermine how we will reproduce ourselves. That is, it seems unlikely that we have a choice about whether we procreate by mating with members of the opposite sex (as humans routinely do) or by occasionally splitting ourselves in two (as bacteria usually do). But it turns out that even though an animal's reproductive behavior may not be of his or her own choosing, that behavior may not be a matter of genetics or evolution or physiology, either.

Some time ago, entomologists studying wasps and wood lice (which most of us call sow bugs) noticed that some species of these insects reproduce parthenogenetically, that is, without males, in fact without mating, and produce only or mostly female offspring. The entomologists concluded that these wasps and wood lice had evolved this method of reproduction to gain some advantage beyond our current understanding of biology (not to mention pleasure).

The entomologists were wrong. This sort of sexual behavior in wasps and wood lice isn't normal, it's a disease—an infectious disease.

Entomologists leapt to the wrong conclusion because of something they couldn't see—something hidden inside the wasps and the wood lice. *Wolbachia pipientis,* a bacterium and obligate intracellular parasite, lives in the ovaries and testes of many insect species (maybe as many as 16 percent of all insects—some two to five million species—are infected with one strain or another of *Wolbachia*). But the bacterium is only transmitted vertically, from mother to offspring, that is. That means that only female insects can transmit the infection. Because of that, it's in *Wolbachia's* best interest to limit the number of male offspring. To accomplish this, *Wolbachia* has developed a whole series of ways of screwing up its host's sex life.

In wood lice, the bacterium interferes with hormone production and/or the action of those hormones and changes infected male wood lice into female wood lice. In other insects, *Wolbachia* induces a state of "cytoplasmic incompatibility" between males and females which prevents males and females from any productive mating. And in some wasps, *Wolbachia* has completely eliminated males from the species. For these wasps to survive, the females must resort to parthenogenesis, and under these conditions they produce only female offspring. Sixteen percent of all insects, and *Wolbachia* isn't the only bacterium capable of this sort of sex selection by elimination of males or alteration of male behavior. At least five other species of bacteria similarly eliminate males from insect species to accelerate bacterial transmission in the wombs of females.

Males turned into females. Entire species of insects that are sexually inept. Species of insects where males have disappeared altogether. And all of this aberrant behavior, all of these behavioral "disorders," can be "cured" with antibiotics—eliminated completely by any of a number of drugs that destroy bacteria.

Okay. But that's ants and wood lice. Bugs. Mammals are a lot more complex than ants and wood lice, aren't they? Rats are intermediate hosts for another parasite, a single-celled, protozoan parasite called *Toxoplasma gondii. T. gondii* begins and ends its life cycle in domestic cats. The immune response that cats make against this parasite forces the parasite into very tough cysts that are shed in cats' feces. These cysts survive in soil for years, waiting for an intermediate host, in this case a rat, to eat the cysts. Inside of the rats, *T. gondii* resumes its life cycle. The ultimate goal of the parasite is to complete its life cycle by returning to its primary host, the cat. Cats do not have a great fondness for dead animals. So *T. gondii* doesn't kill its rodent host. But rats have a general fear of cats, and avoid the scent of cats and their urine at all costs. This slows the transmission of *T. gondii.* To overcome this bottleneck, the parasite has learned to make rats crazy. Rats infected with *T. gondii* show no fear of cat urine. That isn't because these rats have no sense of smell. Because while some infected rats seem indifferent to cat urine, others develop an attraction to cat urine, an often fatal attraction. Mad mammals, seemingly inviting their own death. Again, no dysfunctional family or birth defect or blow to the head made these animals crazy. This madness is infectious.

T. gondii also infects people. Gardening in cyst-infested soil, handling infected meat, emptying litter boxes used by infected cats can all result in infection. In fact, nearly 50 percent of the people in this world have *T. gondii* cysts in their brains. *T. gondii* has never figured out a way to make humans palatable to cats. But that doesn't mean people are unaffected by this parasite. Women with *T. gondii* cysts in their brains were more outgoing and warm-hearted compared to uninfected controls in psychological tests. And men infected by *T. gondii* were more jealous and suspicious than uninfected men. Behavior with a twist, a protozoal twist.

———

What was it that drove Vincent van Gogh to take his own life? Depression was part of it, certainly. But depression alone seems insufficient to explain all of Vincent's behavior. In the year before his death, van Gogh enthusiastically brought Paul Gaugin to join him in Arles. Less than two months later, he attacked Gaugin with a straight razor, and then, in his remorse, Vincent cut off his own ear and offered it to a local prostitute. In the same year van Gogh painted *Sunflowers,* a celebration of yellows and browns, and *Starry Night,* with its haunting blues and swirling stars—a portrait of the abyss itself. Within a few months of one another, he also painted the inviting *Bedroom at Arles* and a brutal self-portrait which leaves the viewer nowhere else to look. Later that same year, he painted *Starlight over the Rhone* and the crows of *Wheat Field with Crows.*

Those dramatic swings between euphoria and abject despair make it appear that van Gogh suffered from bipolar disorder, that he was manic-depressive. At the height of his mania while painting *Sunflowers.* In the pit of his depression while daubing crows over wheat fields in what would be his last painting.

Bipolar disorder is a "behavioral" disorder, a mental disease, a disorder of the mind. The causes of bipolar disorder are unknown, but many of us might agree that within the bipolar patient, the crows are all inside that person's mind. We might be wrong.

Rats, tree shrews, and monkeys—mammals like us, some very like us—that become infected with Borna disease virus behave much like humans with bipolar disorder. These animals exhibit periods of apparent mania and periods of obvious depression. They are more anxious, less sexually active, less interested in food, and have a greater desire for salt—just like manic-depressive humans. All because of a virus, another obligate intracellular parasite. And because of that virus, infected animals—and only infected animals—develop abnormalities that mimic a devastating behavioral disorder of humans.

Humans are also susceptible to infection by Borna disease virus. And those of us who become infected with this virus appear more likely to develop certain "behavioral" disorders. At autopsy, nucleic acid from Borna disease virus has been found hidden in the brains of a disproportionately high number of people with afflictions such as bipolar disorder, severe depression, and schizophrenia. Viral madness?

Perhaps it was all in Vincent's head, all in his mind. But even if we are unwilling to change our thoughts about what "mind" means, we may have to change our thoughts about what "all" means.

———

Obsessive-compulsive disorder (OCD)—another devastating behavioral disorder—manifests as an inability to resist or stop continuous, abnormal thoughts or fears and/or ritualistic, repetitive, and involuntary behavior. People with OCD may not be able to stop washing their hands, or stop hoarding things, or stop checking to see if they've turned the stove off, or stop driving around the block looking for accidents and their victims. OCD is a mental disorder, experts tell us, a behavioral anomaly.

Unexpectedly, though, a significant number of children who develop obsessive-compulsive disorders develop the first signs of the disease a few weeks after a streptococcal infection—a strep throat. Streptococci are bacteria, infectious bacteria, the same bacteria that cause scarlet fever, rheumatic fever, glomerulonephritis, and other diseases. Apparently, after strep infections our immune systems may mistake our own cells as our enemies. In the case of some people with OCD, the immune system's enemy appears to be part of the brain. These people's immune systems produce antibodies that attack the cells of the brain, and, almost overnight, the "craziness" that we call OCD can develop. And antibiotics, things that kill streptococci, often ameliorate the symptoms.

An infectious disease is amplified somehow by a person's own immune system, and abruptly a person we once called "sane" can't get it out of her head that she is going to harm her own children, can't stop counting the silverware, can't stop scrubbing her hands, can't stop thinking about murdering her husband.

———

And then there's my uncle Henry, the one who disgusted my mother. His cursing, his personal habits, his appearance were odd, I'll admit. All of us, I suppose, imagined Henry was a little crazy. My mother imagined it more than most. And, you remember, she resented his craziness most of all. As it turned out, though, there was more to Henry than met our eyes. Henry had syphilis, an infectious disease that's been around since at least the sixteenth century, when it was named the "Great Pox." A bacterium, a spirochete called *Treponema pallidum,* causes syphilis. *T. pallidum* is transmitted through small breaks in the skin that occur during sexual intercourse. Within a few months, the bacteria spread from the point of entry to the lymphatics, the joints, and the skin. And, in about 10 percent of people who go long enough without treatment, bacteria spread to the brain and spinal cord. Once inside the brain, the spirochetes cause some paralysis and a progressive dementia. Apparently, Henry was among that 10 percent.

By 1941, when Alexander Fleming finally got around to developing penicillin to the point of clinical application, it was too late for Henry. Before that, Paul Ehrlich and others used Salvorsan, a compound that contained, among other things, arsenic. Sometimes it worked, sometimes it didn't. But often the infected understood so little about the disease that if they sought treatment at all it was usually too late. And even now, antibiotics are of little use after the spirochete is in a person's brain. Then there is only an unstoppable, progressive "mental" disease. So for many like Henry, uneducated about the dangers of casual sex and infected before Fleming's penicillin, there was only a final madness.

In spite of that, my mother still resents Henry and, I'm sure, the nature of his infection. I doubt, though, that my mother ever connected the two, ever imagined that the fire beneath the boiler of Henry's craziness was being stoked by bacteria. She never spoke to me of Henry's disease until I was much older. Old enough, I guess, that she thought I would understand even though she never did.

And she never forgave him, never believed that all of it wasn't Henry's fault—the disease, the craziness, the indiscretions. She discounted the disease and laid the blame squarely on Henry. She hated him for that. She hated him, too, I'm sure, for the way Henry's craziness freed him from all responsibility and left her to bear his shame.

Now, as I have said, my mother has Alzheimer's disease. Another disease with no apparent cause. Some people believe there's an infectious agent involved. Maybe there is; maybe there isn't. As her disease progresses, a protein called amyloid is being deposited in my mother's brain and parts of her brain itself are slowly disappearing. At autopsy, the brains from Alzheimer's patients often look like wispy growths of pale coral with deep fissures and frail fins.

Everything that I believed was my mother is slowly yielding to the disease. Her poverty is nearly complete now, her disease nearly crystalline. Her craziness, fulminant.

Now, she rarely recognizes my father—her husband for more than sixty years—or any of her four children. Long-term memory seems more resistant. The mention of my uncle Henry still causes my mother to twist her lip into a sneer and to curse Henry for his craziness.

———

We are told that van Gogh killed himself because he was depressed or crazy or both. But who knows? Who knows what truly tortured this man or what more there was, if anything, beneath the lurid oils and the feverish brush strokes? Regardless, for at least one moment, the picture of Vincent's death seemed less terrifying to him than the picture of his life. And, though he may have tried to prevent it, that dark picture finally crawled out of the end of his brush and onto his canvas. It was all suddenly there in the crows. Vincent saw that. And when he was finished, Vincent stared long and hard at what he had done, then covered the painting and left his studio. A week or so later, he shot himself, perhaps repeatedly. Two days after that, he died.

All goes onward and outward, nothing collapses,
And to die is different from what anyone supposed and luckier.

—WALT WHITMAN

Acorns of Faith

There is a woman kneeling before a headstone. She is whispering to herself and her hands are cupped just above her knees as though to catch her words before they spill into the yellow dust and skitter off like scorpions. It is hot. The afternoon air smells like old newspapers and sheep dung. As the woman prays, the wind works at a paper wrapper pinned beneath her shoe. The wrapper whispers its own prayer. Overhead the sky is an uncompromising blue.

We have driven most of the day to get here. But just now we are waiting inside the broiling VW microbus, draining the last of warm beers. Neither of us has said anything since we arrived. Will sucks at his teeth and pulls a loose thread from a hole in his ragged blue jeans. I watch the woman at prayer.

Suddenly, the wind rips an old photograph from the dashboard of the van and sends it curling out across the Texas plain. The photograph was of a dog, some mixed breed with its tongue hanging from its mouth. It is Will's van, but it wasn't one of his dogs. In fact, Will had no idea whose dog it was or where the picture had come from. The Polaroid dog settles in a scrub-filled arroyo beyond the graveyard, and the wind goes back to working the yellow soil, the same soil that wind has been working every day of every month for a million years.

At last the woman's lips stutter to a halt. She draws a long breath, releases it, then brushes the words from her hands into the soil. She rises, grabs the sides of her dusty skirt against the wind and walks to her car, an old Toyota, red paint peeling from the hood, one mirror hanging by a thread. She glances at us as she passes, but we don't register inside her dark eyes.

"Hubba-hubba," Will mutters, eyeing her as she passes through the rearview mirror.

The woman slides onto the vinyl of her car's seat and pokes an arm through the open window. She doesn't look back. Her black hair spills from the Toyota as she drives off. The dust her car kicks up rolls through the windows of the old van, muddies our nostrils and cakes our lips. Will wrenches open his rusted door and gets out. I follow.

In front of the van, there is a gate but no fence. Beyond the iron gate, there are incomplete rows of oak trees. Massive, out of place here. Rows like brown teeth, with every third or fourth one kicked out. Behind each tree, there is a marker—wooden or stone—a foot or two high. On each marker there is, or was, a name.

"This is it," Will says, "like I told you."

I can't think of anything to say.

We wander through the cemetery for an hour or more, looking at the trees, trying to read the names on the stones—the wooden markers were long ago scrubbed white by the sun and the wind. Will squats before a treeless stone. I stop beside a bit of granite with the name *Pearl* chiseled in it. My grandmother's name was Pearl. I stretch my arms around the tree that rises in front of Pearl's stone. My fingers don't quite touch.

Will steps up behind me.

"Incredible, isn't it?"

"Yeah, it is," I say.

And incredible, I think to myself as I release the tree, *is exactly right.* Each of those who fought this god-awful land for a living, laid at the end—while a few skinny neighbors in overalls sang

"Amazing Grace"—into one of these shallow graves with an acorn beneath his or her stiff and callused hands. And now, where ribs and breasts once lay, oaks have busted up through this old Earth and stand whispering to one another in this old wind.

I put my arms around the tree again. The bark feels rough against my too-soft hands. Where the wood has split from the near constant drought here, the tree smells like bone.

"Let's go get some *cold* beer," Will says.

"Okay."

I give Pearl a final squeeze. The wind sends a curl of dust dancing among the oaks. The sky is still without compromise.

Will closes the gate, then steps around it, and we climb back into the van. Doors slam, wind and steel-against-steel, then silence.

"Pretty amazing, huh? These old gomers worked this land for years and all that's left is those oak trees. There used to be a town somewhere near here. Down that road over there, I think. But it was gone before I ever came here. Covered over or ground down to nothing, I guess, by the caliche and the goddamn wind. Nothing much between here and Colorado to slow down that wind. Maybe being buried with a fistful of acorn isn't so bad."

He cranks the old engine over, then punches a button and Jerry Jeff Walker blares inside the cab. Dust swirls up and into the van. We crank the windows shut and the heat grows fierce.

———

As I watch the caliche uncoil behind the van, I realize I am unsettled by what I've seen, addled by such an obscene show of human faith. Mostly I stand inside of spaces where faith doesn't hold much sway, places where people with faith are asked to be quiet about it. But here, everything is nailed in place by faith, everything is red with hope. Men and women who left log or brick homes in Europe or New England for sod huts in the middle of this prairie, droughts and plagues of grasshoppers that took everything but the dirt, stillborn children and dust storms that blackened July noons dark as coal mines. Livestock that dried up like figs or swelled up with the bloat. Women who, in spite of it all, bore children out here, then somehow found ways to give those children enough hope to gather up the tools their parents had left them and press on. And after all of that, all of those brutal days and nights, these people wished for nothing more than to be raised up once again into the same blistering sun and burning wind that ruined them once before. Nails, big as oaks. And if it weren't for those nails, all of it, even the brittle land itself, would have blown away.

A slab of Texas slides past the window, flat as a skillet. Turkey buzzards work the thermals overhead. Jerry Jeff sings of the Hill Country rain. Will plays along on his harmonica as he drives. I'd been to Texas before—Dallas, Houston, places like that—arriving and leaving by air. But this is the first time I've been cross-country like this, first time down the little, sometimes unpaved roads and through the Texas backwaters. There's a nearly dazing monotony to it all.

Women who watched scarlet fever and chicken pox eat their way through families. Men beaten blue-black by their fathers and abandoned by their sons. What the hell right do these people have to faith like that?

We finally find a paved road. I roll down my window, and the dust clears from inside the van. Outside, some unremarkable county spreads out beneath our wheels. A bluebottle fly works its way across the inside of the windshield, startled at the sudden solidification of the universe. The fly looks longingly at all that lies beyond the windshield, though it isn't much—oil derricks and grain silos being the only things that interrupt the view much between here and the Gulf Coast. I gather the fly in my hand. It squirms, imagining, I guess, this is the final crossing. It says its fly prayers, working its forelimbs frantically, it hums its fly psalms, and then it grows silent, awaiting the inevitable. I flip it out the window as Will slows for a gas station with a Shiner sign glowing in the window.

A man with a ball cap advertising Red Man chewing tobacco is pumping gas into a Ford pickup at the pump next to us. His pickup looks like it might have survived a West Texas rock storm—barely. He smiles at me.

"Howdy," I offer, somehow suddenly caught up in this Texas thing.

"Howdy," he offers back. "Pretty nice day."

"You think?" I ask.

He pumps gas for a bit, then raggedly spits a long dark stain onto the asphalt.

"Around here, I figure any day that my truck doesn't boil over or most of the dirt isn't on its way somewhere else is a good day."

I can't help but notice that most of the truck's exhaust system is tied on with bailing wire and that one headlight is staring fixedly at the parking lot pavement as though looking for a lost contact.

"I see," I say. And then, still trying to pretend that I under-stand what's going on here, "Nice truck."

"Thank ya. Ya'all take care now," he suggests and steps up into his truck bed to push some sacks of fertilizer toward the cab.

Will climbs back into our van from the opposite side with a cold six-pack of Shiner.

"Man," he says, "I'll be surprised if the guy next to us makes it out of the parking lot here."

"Me, too," I reply. "Me, too."

We turn east onto the highway, the sun turns with us, and Will opens two beers. Some part of Texas that looks just like every other part of Texas rolls beneath our van. Will swerves to avoid an armadillo. "Sturdiest damn mammal alive," he mutters.

"They're so myopic they wouldn't survive otherwise. But as it is the automobile is the armadillo's only natural enemy."

"No shit?"

"No shit."

———

What the hell right do these people have to faith like that? I try to compose a list of all the things I have faith in. I even get out my notebook and start writing some of it down.

"What're you doing?" Will asks.

"None of your goddamn business," I reply cordially.

"Oh."

I have faith in . . . , my notebook entry begins, then ends. Everything that comes to mind right off isn't something I have real faith in. They're just things I depend on and believe in but don't really understand. I believe that when I turn on my radio and push the appropriate buttons music will come out of the speakers. But since any one of several engineers could tell me where the music comes from and could, if asked, build a radio from the everyday magic of transistors and diodes, I do not count this belief in something I don't understand as faith.

Birth control, televisions, pickup trucks, the stock market, potato peelers, computers, ballpoint pens, supermarkets, light-bulbs, digital watches, the moon, and the stars fall by the faith wayside as well. I depend on these things, I rely on these things, and I expect certain behaviors from these things, even though I have no idea what the bases of these behaviors are. But I do not have faith in these things.

Instead, I write in my notebook, *Faith is a belief in something you have absolutely no reason or right to believe in.* Mentally, I list: faith in God, faith in the church, faith in your spouse, faith in the next sunrise, faith in the motions of the stars themselves.

Then I write: *Faith is beyond thought. Faith is an absolute certainty of something that is so patently absurd you could never justify it to anyone who didn't already have the same faith.*

That's faith. *Faith is burying someone and not just leaving the body in the rocks where it fell. Faith is a compass. Faith is planting grain. Faith is leaving. Faith is folding an acorn beneath the broken fingers of your dead wife's hands.*

————

And then I write: *Faith is putting on clean underwear.* Will hands me another beer, still cool. I sheathe my pen and turn my eyes back to the road.

Will then tells me again how the Shiner Brewery was the only brewery in the country that continued to brew beer right on through the entirety of Prohibition. The gaps he leaves in the story, I fill with information gathered from the past two hundred times he has told me that same story. It's one he enjoys.

In its effort to make the blue reach from level horizon to level horizon, the sky is stretched almost too tight here. It looks pained. Turkey buzzards are skimming the highway for carrion, and a meadowlark's cry cuts dryly through the clear air. Texas is a place of stark contrasts—meadowlarks and turkey buzzards, Pentecostal revivals and David Koresh burnings.

————

I was raised as a scientist. Not by my parents, of course; they were Catholics. Devout—believed Mary was a virgin, Christ was the son of God, masturbation or missing mass on Sunday would land you in hell for all eternity. More or less, I raised myself as a scientist. My brother Michael helped when he could. But when I was thirteen, Michael left to prepare for the Catholic priesthood—a thing he later got over. So mostly, my scientific upbringing was up to me, which might be viewed by some as disadvantageous. I never felt that way. In fact, for the most part, the way things were suited me just fine.

But there is a down side to growing up a scientist.

Science shows you the inside of things, the parts most people never see. And once you've seen the inside of things, the outside of things never looks quite the same. Science raises issues that make you stop and think—issues like thymuses and parasites that make people crazy, issues like the character of self and chemistry of fear, things like orchids and syphilis. And science gives to the world a certain symmetry that, real or not, I find pleasing.

However, science as a discipline gives short shrift to the concept of faith. Because of that, when confronted with something like what I saw in that graveyard this morning, I'm at a bit of a disadvantage.

———

Suddenly, the beer bottle slips from Will's hand. As he's fishing around for it near the gas pedal, his head beneath the van's dashboard, the van begins to wander onto the gravel berm at the roadside. I reach, take the wheel from Will's hand, and move the van slowly back onto the highway. Just as the van's rear wheels fully reach the pavement, the berm disappears, and in its place a ravine opens, cleaves deeply into the hard, dry earth. As we flash past, I notice a coyote feeding on something in the bottom of the ravine. Will locates the beer bottle, rises from under the dash, and downs the last of his beer.

"Thanks," he says, then belches, takes the wheel back, and tosses the empty Shiner bottle into the back of the van.

Did I mention Will's a scientist, too?

———

Will and I met in a lab in southern California. Both of us had come to learn immunology. The day we met, Will was trying to steal a desk I'd laid claim to a week earlier—this desk was solid and more or less fixed in place. The one next to it was on wheels (something about access to a utility corridor). At the end of the day, thanks to one of the other scientists working in that lab, one of the nicest men either of us would ever meet, I had the desk back and Will had wheels. By the end of that week, I thought Will was one of the most remarkable people I'd ever come across. I still do. An odd mix of Woody Guthrie and Elie Metchnikoff, Hunter Thompson and J. Robert Oppenheimer, Stephen Dedalus and Martin Arrowsmith. That was about 1974. For the next five years or so, Will and I spent a lot of time together.

Within a year after we met, Will and I were at the very front of what was going on in immunology. Especially Will. He was the centerpiece of our lab, and that lab was one of only two in the world working to put the pieces of self on the table. It was Will's job to put together the chemistry that would show us all just how human bodies figured out what belonged and what didn't. I was trying, with much less success, to follow Will's lead using mice instead of humans. Will had the best pair of hands I'd ever seen in the laboratory. Everything he touched worked. He was smart, funny, and seemed to know most everything there was that was worth knowing. When I met him, I was alone in that city, didn't know a soul. Will and his wife changed that. Over the next two years, the three of us had more fun together than I thought possible for just three people. Almost everything we did, we did together. It was the science, though, that held Will and me most closely. We spent nights working on motor-cycles or drinking beer or camping under palm thickets in Mex-ico and always, always we were arguing immunology. It was a good time. But between the science and the fun, there was no time to think about faith.

I think it was Plato who first proposed that as the crystalline spheres of heaven rotated concentrically around one another, they produced a heavenly music, the song of the cosmos itself—the "music of the spheres." Of course, no human has ever noticed that music, because we've all been listening to it since birth. And, like other things with us since birth (the rotation of the Earth, for example), we have no sense of it, no ear for it. We would only notice if the universe suddenly ground to a halt and the music of the heavens ceased. Only in the sudden silence that followed would we hear the final strains of the tune. Like when the furnace kicks off, and you suddenly hear the echoes of the fan—the noise your brain was tuning out until the blower shut down and the fan's blades ceased spinning. And because we've never heard the music, we assume that everything will be fine, the music will go on playing forever and everything will be fine. That's faith, whether we speak of it or not.

Every day, every one of us inhales a billion or more living things—viruses, bacteria, funguses. Yet not one person I know expects to be sickened by this, in spite of the fact that almost no one I know understands how a human's immune system works, how bacteria or viruses do what they do, or much cares. The music has been roaring in our ears since birth, so no one expects it will stop.

But in 1918, the music stopped, completely. Before it began again, six hundred thousand Americans—more men, women, and children than were killed in World War I, World War II, the Korean War, and the war in Vietnam *combined*—lay dead. And across this sweet planet, while we waited for the music to begin again, thirty million more people died.

The virus probably began in ducks and was passed to a pig when an infected duck—flying across the Kansas plains—defecated in a pig's trough or drinking water. Then a pig passed the flu to humans when the pig sneezed in a little girl's face. The little girl shared it with her family. The family went to school and church, where they shared the flu with the faithful during church socials and gospel hymns. Then U.S. troop trains and ships spread the flu throughout Europe and from there, to the world.

Thirty-three thousand, three hundred and eighty-seven dead in New York City alone—mostly twenty to thirty years old, mostly healthy, mostly with a whole life to live—all dead.

When it was over, the dead were buried, the living picked up where they left off, and everyone got on with their lives.

For a year or two after 1918, people could hear the music. For a year or two, they knew that each breath they drew might be their last. They knew, too—whether they still heard it or not—they were dancing to a complex tune, and that tune made them sad and nervous. But as the music played on, it fell beneath the hum of American industry. The living stopped wearing surgical masks, stopped glancing nervously over their shoulders, and eventually forgot all about it.

That's faith, too, whether we speak of it or not.

———

We slow as we pass through one more small town in a seemingly endless string of small Texas towns. But something's different here, something's happening up ahead. A couple of cop cars and a fire truck are stopped in the middle of the road, lights all going. We creep closer. People line the street.

"What the hell?" Will says, as we nose our way past the Woolworth's store.

Then we slide beyond the blistered facade of the Great Western Hotel and turn onto Main Street and it becomes apparent that we have inadvertently wandered into some sort of celebration. Whatever it is, the citizenry is out to celebrate. A few tattered banners snap in the dry breeze, dogs and kids are everywhere, ice cream and watermelon. What they're celebrating is anyone's guess.

The police barricades force us to detour around the town center. Will moves his beer from the dashboard to the floor of the van. "Goddamn gomers," he curses only half seriously, and we move off down a tree-lined side street. The whole town smells of barbecue sauce and roast beef. Everyone here seems to know everyone else, and everyone seems to think everyone else is making an even bigger fool of himself.

Yards are mostly dry grass and dirt. An occasional house stretches to two stories. Here and there, there's a coat of fresh paint. At the edge of town, there's a park, an acre of brown grass, a swing set, a slide, something that might be a ball field, and some picnic tables. The tables have been covered with butcher paper, watermelons, and paper plates. Will and I make a quick left, back onto the main highway, and roll out of town.

"God," Will sighs, "small-town Texas."

The Lone Star State spools out in front of us, indifferent.

———

In 1918, people were no more vulnerable to influenza than we are now. In 1918, treatments for flu were not a lot worse than they are now. And nothing we know about the flu of 1918 sets it apart from this winter's flu, or that of next winter, or the one after that. It's been over eighty years since the last person died of Spanish flu. But we haven't changed since then, and antiviral therapies have changed only minimally. Nothing much has changed since 1918—except *influenza*. Influenza's changing all the time.

" 'In one place after another pestilences.' (Luke 21:11) Right after World War I, some twenty-one million people died of the Spanish flu. *Science Digest* reported: 'In all history there had

been no sterner, swifter visitation of death.' " That from the Je-
hovah's Witnesses who take the Spanish flu as one more sign of
impending world doom. Who knows?

If it happened again tomorrow, the Spanish flu might kill sixty
million people, maybe a hundred million. But almost no one (me
included) believes it will. I write that in my notebook. *The Spanish flu won't happen again tomorrow.*

Faith.

Every breath we draw is filled with possibilities—some of
them wonderful, some of them horrible. But we go on breathing,
and most of us never give it one thought. The *Sin Nombre* strain
of Hantavirus has killed more than ninety people in Utah and
New Mexico. Dengue fever, shipped in by mosquitoes, killed
thirteen in Texas, and worldwide dengue virus now infects a hundred million people—90 percent of whom are under the age of
fifteen, and 24,000 of whom will die from dengue fever. Since
1995, five hundred cases of monkey pox have been reported in
the Congo. Most of these were in children under the age of
sixteen. Many epidemiologists think that monkey pox will be the
"scourge" of the new millennium, as bad in its time as smallpox
was in the previous millennium. Over two hundred people sickened and seventy-two killed in Uganda in the most recent out-
break of Ebola virus in Africa. West Nile virus has sickened sixty
people and killed seven in New York City. In 2000, a new ar-
enavirus, named the Whitewater Arroyo Virus, killed at least one
fourteen-year-old girl and probably two others in California.
Somewhere between one-fourth and one-third of the world's population is already infected by malaria. Presently, most of these
people live in the equatorial regions of the Earth. But as the globe
warms, the parasite is moving north. Last year 15,000 people
were bitten by ticks and developed Lyme disease. Somewhere
between twenty and forty million people are currently infected
with HIV, and irresistible sex between consenting adults is one

of the very best ways to spread the disease even further. Still, we all go on breathing in and slapping at mosquitoes, pulling off ticks and sweeping the mouse dung from our garages, putting on clean underwear, taking it off again, making love, having children, and planning for next year and the year after.

Ebola virus, African green monkey virus, anthrax, tuberculosis, Rift Valley fever, spongiform encephalopathies, lepromatous leprosy, hepatitis B, necrotizing fasciitis, tuberculosis, Epstein-Barr virus, cytomegalovirus, malaria, Chagas' disease, leishmaniasis, babesiosis, plague.

Trypanosomiasis, schistosomiasis, toxoplasmosis, yellow fever, histoplasmosis, encephalitis, meningitis, cryptosporidiosis, onchocerciasis, dracunculiasis, typhus, tularemia . . .

————

It might seem simple, sometimes even appropriate, to substitute "foolish" for "faithful." At times, I have. Faith *is* inherently foolish. I write that down. *Faith is foolish. Faith is something we've no right to.* But, even as I write the words, I know that's only part of it.

To test my hypothesis, I ask Will.

"Will," I say, "is faith foolish?"

"Huh?"

"Faith. You know, your beliefs."

"Oh. No, goddammit. I have a powerful faith in my beliefs. Right now, I believe I'll have another beer." And he does.

It's growing dark outside. The lights of Texas's one hundred million or so small towns are beginning to glaze the flatlands like handfuls of tourmalines and sapphires tossed by some bored oilman. Will sets his beer between his legs and starts fiddling with the radio, rolling the knob across the dial. For a moment it's a pastiche of Tanya Tucker and Tammy Faye, George Jones and

Billy Graham, "He Just Stopped Lovin' Her Today" and the Sermon on the Mount, "Stand by Your Man" and the Twenty-third Psalm. A land of stark contrasts.

Finally, Will finds what he's looking for and turns up the volume. It's Reverend Ike. And as the Texas night gathers in the gutters alongside the roadway, Ike explains how "You can *have* what you wanna have. You can *be* what you wanna be. Just have faith, sister, and send *me* your money. Trust in Ike and the Lord to set you free." This is followed by series of testimonials. Earlene from Bastrop explains how "I prayed to the Lord, I sent my money to Reverend Ike, and the very next day in my driveway there was a brand-new Chevrolet Monte Carlo convertible, maroon with dove-white leather seats and a stereo. Praise the Lord." Bobby from Dallas, Rayette from Del Rio, Floyd from Fort Worth, and Charlene from San Angelo. Each story as stirring as Earlene's. Each moving Ike a little closer to God. I think of Janis Joplin and her plea for a Mercedes-Benz.

As if reading my thoughts, Will starts playing a bluesy version of some Alvin Crow tune on his harmonica, and the stars come out in full evening dress to watch. It smells like rain.

———

Orchids have evolved intimate relationships with insects. One species of orchid must be pollinated by a wasp. For this to occur, the wasp must mistake the orchid for a female wasp and attempt to mate with the plant. Imagine that.

The orchid, for its part, must believe that, at the right moment, there will be a wasp. The wasp must believe that he is a remarkably handsome fellow, and that there is no better time than right now for the creation of more wasps. Both must believe that sexuality in any form justifies tomorrow.

The males of at least one species of wood louse spend their

entire lives inside the bodies of female wood lice. Their only role in life is to fertilize the female's eggs. From those fertilized eggs will come another generation of wood lice—each of them female, each of them already supplied by the louse god with a single male embryo. And the very first task for these female wood lice is to consume their brothers and chew their way out of their mothers, just as they will one day be chewed to pieces by their own children—children already growing in each of their wombs. And yet, female wood lice go on producing eggs as though being eaten by your children is a small price to pay for having children, and male wood lice go on behaving as though a life of darkness is a pretty good life.

The replication of chromosomes, the turn of the rotor that drives a bacterium, the first convulsion of a fetal heart, the rush of sea turtle young into the ocean, ejaculation. The simple act of cell division—the simple belief that there is enough for two.

Things we've no reason to expect or to believe in.

Zinc, copper, cobalt, selenium, iodine, and the rest of the periodic table are the rivets that bolt us to the living. Without any one of these, human life is impossible. Yet none of these chemicals originated on Earth. It takes a stellar furnace to fuse enough hydrogen to make cobalt or copper or zinc. But neither did any of these come from our sun. That star's furnace doesn't burn hot enough. Copper and selenium and iodine and cobalt and zinc are only created on the white-hot anvils of supernovae. Only as great stars die does the forge get hot enough and the smithy's arm strong enough for hammering hydrogen into any one of these. Red giant, to white dwarf, to black-hot womb.

Only in the clenched fists of these dying stars do pressures, driven by gravity itself, reach levels sufficient to gather the seeds of hydrogen and squeeze protons and neutrons into the stuff of life. Phosphorus, iron, oxygen, nickel, zinc. Each of us—clam, cricket, primate—an elemental acorn, carried, at the end, inside a star's wizened fist. Each of our elements risen from another's grave. Each of us, the last great wish of a dying star.

Something we've done nothing to deserve. Something nearly absurd.

———

At Will's place, we roll out of the van kicking beer bottles everywhere. Will's wife, Sarah, watches from the doorway. She smiles, having seen this a time or two before. It's an old home, overhung by ancient pines. Will and I gather up the beer bottles and toss them with much ado into an aluminum garbage can at the rear of the house. I smell honeysuckle.

Sarah has made enchiladas. While she heats them, Will pulls a bottle of tequila from an old, black-oak cupboard. He cuts some lime and grabs two shot glasses from above the sink. We sit.

"So, what do you think of Texas?" Will asks as he pours tequila.

"I was surprised to find trees," I say.

"Seriously," Will demands.

"I liked the graveyard. I like Austin. You can keep the rest."

"Y'all went up to that cemetery today?" Sarah asks, spreading a thick sauce the color of old brick across the top of her enchiladas. The room reeks of cumin and roast chiles. John Prine's "Angel from Montgomery" is playing low in the living room.

"We did," I reply, take a sip of tequila, and reach for a lime.

"Pretty amazing, isn't it?" she asks.

"It is," I say. "One of the strangest places I've ever been."

"Stranger even than that old Mercedes graveyard we used to visit out by Tecate," Sarah remarks, smiling at me.

"I agree."

"I've still got them car doors somewhere," Will says. Then, "I gotta pee," and he heads off to the bathroom.

Sarah and Will and I have spent a lot of days and nights together, camping, drinking, sometimes crying. Sarah is a short, dark-haired, full-breasted woman. I watch her working the enchiladas. There is something about Sarah that tends to make men hopeful.

"How are you doing, Sarah?" I ask.

"I'm doing fine," she responds, maybe a little too quickly.

"You sure?"

"I'm sure."

"Sorry we got back so late."

Sarah says, "I don't mind."

I say, "Okay." And I guess maybe it is. Though I'm reminded of the time Will and I took off for the desert one night about midnight so we could watch the sun come up over the desert. We were drunk and didn't tell anyone we were going. Next morning about nine, after we'd slept through the sunrise, Will called Sarah from a pay phone. When he got back in the car, he described the conversation.

Will: "Don't worry, dear, I'm all right."

Sarah: "You wanna bet?"

Will: "Okay. Here's the deal. You can be as mad at me as you want to be, until I get home."

"And?" she says.

"And I promise I won't do this again, for a while."

Sarah: "You just come home."

Will: "Okay."

The woman I was living with at the time took a somewhat different posture regarding my absence, a posture that is difficult to describe if you haven't seen it.

And I'm reminded of a time Sarah dragged me, three-quarters

unconscious, out of bed at about three o'clock in the morning, wanting to know where in the hell Will was. He and I had left a local bar together, but before we got to his car, a woman had diverted his attention, among other things, and he had left with her. I drove his (actually Sarah's) Volkswagen van back to my place and went to bed. At that moment, half-clothed and completely confused, I could think of nothing but the truth to tell her. So I did.

She heaved up a complex sigh and said, "I'm taking the van and going home." She'd come in his car.

"Okay," was all I could think of to say.

She left. I went back to bed. Next day when they came over to get Will's car, he told me he was half a block from my house when Sarah found him. The girl he'd left the bar with was undergoing a sobriety test. Sarah stopped. Will climbed into her car. The cop glanced at Will, then at Sarah. The cop shook his head, then nodded to Will, and he and Sarah left.

Laughing, Sarah said to me, "That's right. I came around that corner and there was Will—standin' on his thing."

Will, standing next to her that day, grinned sheepishly. "Yep, standing on my thing," he said. We never spoke about it again.

"You want a beer with dinner?" Sarah asks me as she's spooning rice into a large earthenware bowl.

"Sure," I say. "I'll take a beer." And then I ask, "Do you remember what you told me after the night you drug me out of bed looking for Will?"

"You mean about finding him standing on his thing?"

"No, the other part."

"No, I don't," Sarah said.

"I asked you why you put up with shit like that. And you said to me that whatever Will did, you could live with it so long as he came home to you when he was through doing it."

"I'd forgotten that."

"I haven't," I said.

Will returns from the bathroom, drops one last shot of tequila in his mouth, does his best imitation of Jack Nicholson in *Easy Rider*—"Nit, nit, nit," pumping his arm like a chicken wing. Then he sits to eat.

Sarah places the enchiladas, chicken, on the table.

"I want you to hold the chicken," Will says, completing the Nicholson medley.

"You want me to hold the chicken?" Sarah asks dutifully.

"I want you to hold it between your knees."

Will and I laugh. Sarah smiles at Will and serves up the enchiladas. The last of the evening quickly disappears into our mouths.

Things we have no right to.

––––––

In *Fear and Trembling*, Søren Kierkegaard considers at length the faith of Abraham, and in particular the faith that took Abraham to the mountains of Moriah to kill his son Isaac. A heinous crime by most standards—in Christian and Jewish lore, an act of the utmost faith.

The story is told in chapter 22 of Genesis. God commands Abraham to take his eldest son Isaac to the mountains of Moriah and there to slay him. God offers Abraham no reason. Without a word to his wife Sarah, Abraham loads himself and his son onto asses and together they set off for the mountains. Sarah watches from the window as the man and her boy ride down the valley until she can no longer see them. We are told nothing more about Sarah.

On the mountain, Abraham takes hold of Isaac and throws him to the ground, saying: "Foolish boy, do you believe I am your father? I am an idolater. Do you believe this is God's command? No, it is my own desire."

Isaac pleads: "God in heaven have mercy on me, God of Abraham have mercy on me; if I have no father on Earth then be Thou my father."

To himself Abraham thanks God, saying: "It is after all better that he believe I am a monster than that he lose faith in Thee."

Abraham builds the sacrificial fire, ties Isaac's hands and feet, and draws his knife. At the last moment, God stays Abraham's hand and offers a ram for the sacrificial fire instead. Abraham unties Isaac, slaughters the ram, and burns the ram's remains for his God. Then the man and the boy return home. We are told nothing more about the ram.

Faith carried Abraham with his eldest son to that mountain, and there faith led Abraham to believe that the murder of Isaac would please God. And it was faith that whispered to Abraham that God's pleasure would sanctify Abraham's crime.

But what else was left for Abraham as he imagined the look in Sarah's eyes when he returned to her without Isaac? Beneath the boot of an angry God, what else but faith?

Beyond thought, overtly insane.

But the faith of the scientist may be even greater.

"Anybody who has been seriously engaged in scientific work of any kind realizes that over the entrance to the temple of science are written the words: 'Ye must have faith. It is a quality that the scientist cannot dispense with.'" Max Planck, the father of quantum mechanics, said that. Quantum mechanics has taken physicists into realms even Lewis Carroll couldn't have imagined. But you don't need to see the madness of quantum mechanics to understand that Max knew what he was speaking about. How else can you explain scientists' certainty that this universe is orderly and knowable, their unshakable belief that we will one day understand something as infinitely complicated as our own immune systems when we don't yet understand the character of the atom, or our own confidence that their tinkering with the parts of something so powerful and so intricate will inevitably lead to greater benefit and understanding?

Near the end of *Fear and Trembling*, finally face to face with Abraham's unmistakable act, Kierkegaard writes:

> A tragic hero can become a human being by his own strength, but not the knight of faith. When a person sets out on the tragic hero's admittedly hard path there are many who could lend him advice; but he who walks the narrow path of faith no one can advise, no one understand. Faith is a marvel, and yet no human being is excluded from it, for that in which all life is united is passion and faith is passion.

He's right of course, Kierkegaard, but only partly. Passion is faith, but most faith isn't passion. The greatest acts of faith, unlike Abraham's, go mostly unnoticed—unseen by those of us who watch, overlooked by those who act.

———

Every cell whose job it is to defend us from infection arrives in its final functional state as the result of random processes that occur in the bone marrow and in the thymus. *Random* processes. Nothing about these processes guarantees that we will have our biggest guns when we need them. Nothing ensures that we will be prepared to deal with measles when it arrives, or influenza, or Ebola, or tetanus. Yet most of us do deal with them, most of the time. And even more remarkably, nearly all of us believe we will be prepared when the infectious moment comes, that we will not be the ones to die from a friend's sneeze or a tainted hamburger. Not *us*, not *our* children.

Faith.

Nothing about the way our immune systems work guarantees us one more moment on this blue planet, but no one believes that he or she will be the next to fail the challenge and succumb. Every day, every breath we draw reaffirms that faith.

Without such faith there would be nothing in our futures except fear—of a touch, a kiss, a moment outdoors, mosquitoes and ticks, worms and blackflies. And no one with such fear would go on as we do—with our lovemaking, our children, drawing breath after breath. No one.

So, maybe faith isn't the luxury we've imagined it to be. Maybe it is a necessity, as essential as, say, eyeballs and bones, for the survival of humans and others. If that's the case, then we might expect to find faith—like most other essential human traits—buried somewhere in our genes.

Inside of mice there is a gene called *Peg3*. Female mice that have suffered mutations in *Peg3* will not nurture their babies. Mouse pups are always born blind, deaf, and immobile. Normally, the mother is attentive, nurturing. Normally, the mother builds a nest for her young, gathers them into the nest, warms them by standing or lying over them, and nurses them. Female mice with mutant *Peg3* genes do none of this, or do none of it well enough to keep their pups alive.

The protein product of the *Peg3* gene appears to be a zinc-finger protein—a protein that binds to DNA and controls the expression of many types of genes.

Mutations in another gene, a gene called *Mest,* have similar effects in mice. Mothers with mutated *Mest* genes also fail to nurture their young. The protein product of the *Mest* gene is expressed in the hypothalamus—a part of the brain that is intimately involved with behavior and perception, with depression and elation, with immunity and fertility.

Scientists describe *Peg3* and *Mest* as genes that regulate maternal behavior. I wonder if there isn't more to it than that. I wonder if some of what we call faith isn't right there in our genes, the same genes that pushed our ancestors from the sea onto the land, the same genes that pulled our grandmothers and our grandfathers from inside of their secure homes and shoved them onto the plains of North America, the same genes that placed these people's hands in ours while we stared at an irrepressible sunset over the flat fist of an unforgiving land.

Is there any greater act of faith than nurturing children, nurturing them in spite of the unwashed face of this hard world? Nurturing mice may take an even greater act of faith. Mice are fodder for everything from hawks to wolves, and under the best of circumstances live no more than a few years. But nurturing children in a world where infectious disease is still the leading cause of death among children, a world where bacteria outnumber us humans by somewhere around 10^{20} to one, is an act worthy of our most serious admiration as well.

———

Will and I have retired to the front porch. Will and Sarah's dogs—an odd mix of nearly all that is undesirable in the species—are asleep on the sofa, and Sarah, gracious as ever, is cleaning up after us. We're sipping a last bit of tequila, watching the wind in the elm trees and looking for the stars overhead. It's a fine night. A warm breeze, filled up with the flavors of dinners from all up and down the block, works its way across Will's porch. It would be a perfect night except for one little thing. The light from a streetlamp near Will's house douses all but the brightest of the stars overhead.

"Shame about that streetlight," I say. "It really ruins the stargazing."

"Just a minute," Will says, and walks back into the house.

I imagine I've offended him, but shortly he's back—with a gun.

"I got my piece," he says, and shows me the black length of the Daisy Sharpshooter BB gun. He then pumps a round into the chamber, takes a sip of tequila, and draws a bead on the streetlight. A soft pop indicates the gun has fired, and then moments later a faint plink indicates that Will's BB has found its mark. But of course, nothing happens. The BB drops harmlessly to the street.

"Night after night," he says. "It's always the same. You want to try?"

"Sure." I take the rifle, fortify myself with tequila, and fire off a round. Nothing. Not even a plink.

"Let me try another," I ask.

"Okay, but be careful of the recoil. You could put your eye out." The gun, of course, doesn't generate sufficient recoil to displace a bad smell.

"I'll be careful," I promise.

I take one more sip of tequila, draw down on the streetlight, take a deep breath, and squeeze. A soft pop, a faint plink, and nothing. I fire off one more round and strike the light's globe resoundingly. But, of course, without effect.

Resigned, I settle back into my chair and reach for my glass.

Will takes the rifle from me and methodically empties the chamber into the streetlight. Then he, too, settles into his chair and reaches for his tequila.

"You know," he says, "it ain't the magnitude of his victories that makes a man worthy, it's the greatness of his enemies."

"I suppose," I say. "All the same, I wish I could see the stars."

Forgiving the Father

Gray and skinny, the cat had crawled atop the engine during the night—up from beneath, for the heat of the old Ford's rust-red block. It was a cold night. And when the tom finally dozed, it leaned against the valve covers and stretched across the fan belt—just far enough from the black radiator so that it was hot, but not too hot. The chill fled, and the cat dreamt of pickled herrings and soured goat's milk. The warmth of the black Ford wrapped around it.

At six the next morning, my sister Pat shoved a brass key into the Ford's ignition slot and cranked the engine over. The unsuspecting cat screamed as it was first wound and then slowly stretched between the generator and the cam-shaft pulleys. For an instant too long, my sister couldn't place the scream—fan belt, pulley bearing, frozen distributor rotor? Then the sound took shape.

My sister pulled the key from the ignition, sat for a silent instant, and then went back into the house. My father knew what sound the car had made, knew it right away. He'd heard it before.

"Shit," he said out loud, and rose from his oatmeal and walked to the garage.

"Shit," he said once more, as he grabbed his greasy leather gloves, a gunnysack, and a pipe wrench from his wooden workbench. My sister huddled with my mother. I watched from behind the small glass in the kitchen door.

Dad set the pipe wrench on the concrete drive, lifted the bullet-nosed hood of the Ford, and reached in for the cat. What he pulled from inside that old car resembled almost not at all the gray tomcat that had crawled onto the engine the night before. Most of the skin was gone from the cat's face, one leg was hanging from red threads of gristle, bone glistened pearl-white in the morning sun, and the tom's tail was kinked at acute angles in two places. Dad held him by the loose skin gathered at the cat's neck. The cat feebly reached with his foreclaws to protect himself. From where I stood, I could see all of the cat's teeth and most of his jawbone. I could see, as well, the knots in my father's neck and the grim set of his jaw.

Dad let the cat slip from his hand into the mouth of the burlap gunnysack, twisted the top of the bag shut, and lowered it to the concrete. Then he reached for the wrench—red and black, the length of his forearm.

Like a man crushing stone, my father struck at the cat with his wrench. Three, four, five times, before he stopped and opened the bag.

Behind the kitchen door, I couldn't see inside of that bag, but I could see my father as he measured the ruined cat. He looked to me like a man surveying his plowing might look—not proud exactly, but done, and glad to be done.

He looked for a long moment into that bloody burlap bag, then he closed the top and dropped the bag and the cat into a galvanized steel garbage can at the edge of our driveway. To finish what my sister had begun, Dad pulled the garden hose from where it hung on the wall of the house, turned the water on high, and with the chrome nozzle twisted tight, hosed the last of the cat from under the hood of the car. The hose spat ice, then water, and the cat's blood ran red from beneath the car.

When he was done, my father rolled the hose carefully and hung it on its notch by the kitchen door. With the back of one raw hand he made a quick swipe at his eyes, then stepped inside. He glared for a moment at each of us, daring any one of us to speak. He glared longest and hardest at my mother, who'd chosen to keep the cat when it wandered into our yard. Then he sat and finished his oatmeal. And though no one of us spoke, none of us missed his point, either. Certain of that, he picked up his keys and left for work.

I hated him for that, the way he glared at me as he dropped the remains of that old cat into our laps and walked off to work.

———

We all take vows when we are born, each sworn to before the brown prelate of this Earth. If we are fortunate, our first years may make us forget our vows, but those of us who live long enough discover that each of the promises we made must be honored. Mostly, it's poverty we swear to—absolute poverty. We all do that, and we are all bound by it. Forgetting excuses no one.

My father and I were both born into the dead of winter. And born into the long nights of those winters, we, too, took our vows, vows of poverty. But we did more than that, we each swore an oath as well. Even as I lay in my mother's womb, I think, I swore an oath of allegiance to her—an oath that would rise between my father and me. Thirty-five years before me, I'm certain my father swore a similar oath to his mother. And though neither of us had any choice, we took up those oaths like swords and carried them into the dim light of our first winter mornings.

The day the fan belt grabbed hold of the old cat and wouldn't let it go, I recalled the oath I made years before. And while we slurped the last of our oatmeal and glared at one another, inside of me there rose a little hatred for my father. It wasn't the first time I had hated him, only the first that I recall with any clarity. And it wouldn't be the last time that he and I would stand at opposite ends of an iron bridge and throw stones at one another. I have blamed him for that. I have blamed me for that. But perhaps there's more to it. Perhaps the rocks we tossed were handed to us before we could possibly have known what those rocks were for.

———

At the moment of fertilization, we are each given forty-six chromosomes—twenty-three from our mother and twenty-three from our father. Because of that, we imagine our parents' gifts to us to be equal—our father's DNA, our mother's DNA—bestowed in nearly equal amounts upon each of us at fertilization. But the truth is, DNA is not the molecule in charge. On its own, no gene has ever done a single thing, including reproduce. Genotype has never been phenotype. Perhaps we have imagined otherwise because it is men who told us how we came to be. Men contribute little more than DNA to the fetus. Mothers give more of themselves, much more.

Sperm, human sperm, is nothing more than a hairsbreadth of DNA pushed by a tail, a bullet, nearly, sent to pierce an egg—twenty-three chromosomes and an outboard motor.

Eggs aren't like sperm at all. Eggs are cells. Each of these cells holds twenty-three chromosomes as well, but inside of eggs there is much more than just DNA. Inside of eggs there is cytosol—a piece of the sea carried up onto land a billion years ago. Cytosol is water and salt and proteins—things we know nearly nothing about, things that turn genes on, things that turn genes off. Things that determine how and when the whole process begins and which path—of all the paths that might be followed—is chosen by this newly born zygote. The mother always speaks first, and in her words we see our first truths and unlimited possibility.

Among the eddies and swells of the egg's cytosol, there is also the flame—the mitochondria. Each warmth that is ever ours—each fever, each lust, each movement, each breath, every swollen instant—is given to us by our mothers. The flames given to us by our fathers are quickly snuffed by our mothers.

A few of the mitochondria from the sperm pierce the egg. But the egg moves quickly and ejects those mitochondria or destroys them. Each mitochondrion we are left with as children is descended from those given to us by our mothers. The other mitochondria were killed. Our mothers and our fathers may have given us the map, but our mothers gave us the light to read the map by and the tools to extract its secrets.

Then as we study that map and plan our own journeys, our mothers must warn us. They have seen much of the territory that we will cross, and they know the danger there. If we were to set out completely on our own, there is little chance any one of us would survive for even a year. Too many of those who surround us—viruses, bacteria, parasites—wish to have what we have. Before we are delivered into this septic world, we need protection. Biology foresaw that, and tended to it.

While we are still within, our mothers warn us of what sickened them and of where they faltered on their paths, the places where they stumbled and fell and nearly died. Our mothers have to do that, because if they fail to warn us, we will die. And though they might choose otherwise if they could, they must warn us, as well, of our fathers.

Near the end of her sixth month of pregnancy, just as the fetus is making its first sounds, the mother speaks to the child. "Pay attention," she says. "You are in danger. There are things you must know." But the child cannot hear her, it has no ears for her words. So she turns her words into proteins and whispers instead into the umbilicus. She sends her child antibodies, a little of nearly all the proteins that have protected her, the mother, against the microbes that would have killed her if they could.

And in among those antibodies, because she has no choice, the mother sends antibodies that speak to the child of its father. That is as it must be.

Must be, because of the ways that evolution has arranged for our mothers to meet our fathers. In part, perhaps in large part, it was smell that pulled our mothers to our fathers—a smell that swelled within our mothers and opened them to our fathers.

Human chromosome 6 holds a group of genes called the major histocompatibility complex, the MHC. The character of these genes determines many things—whether or not we will keep a kidney transplanted into us from a friend or sibling, whether we will survive an infection by influenza or plague, whether we will develop rheumatoid arthritis or diabetes. The molecules encoded by the genes within the MHC save us, and sometimes destroy us. And in ways we've only begun to appreciate, the molecules produced by the MHC help us choose our mates.

Mice can be bred or changed genetically until the only genetic differences between males and females lie inside of the MHC. Male mice like this, given a choice of mating with MHC-identical or MHC-different females, choose the MHC-different females almost always. And if the male mice are offered instead the urine of females in estrus, in a "Y" maze, the males still seek the end of the "Y" that holds the urine of the MHC-different females much more often than they seek the end of the "Y" with the urine from the MHC-identical female. That is, male mice much prefer to mate with MHC-different female mice, and male mice can identify those MHC-different female mice by the way they smell.

But that's mice. What about humans? Women, too, seem to be able to smell the molecules produced by the MHC. College-age women given the shirts of MHC-identical men find the smells of the men's shirts either unattractive or at least uninteresting. But these same women are aroused by the smells and the shirts of MHC-different men. It seems we all can smell the products of the genes inside the MHC, and when conditions are right, these smells excite us. Millions of years of evolution did that, and it did it for a reason.

For our protection, each of our chromosomes 6 holds six or eight genes inside the MHCs. Since one of those chromosomes 6 was given to us by our fathers and the other by our mothers, if our parents had different MHCs on their chromosomes, we may have as many twelve or sixteen of these MHC genes to protect us. The products of these genes are all we have to recognize and present to our immune systems the rest of the biological world: each parasite, each bacterium, each virus, each fungus, everything that threatens us. These MHC molecules stand between us and the infectious world. It is good to have more than one MHC molecule, because each of them has its own special affinity for some part of the world that would destroy us. Some MHC molecules are better at binding and presenting mumps virus to our immune systems, other molecules are better with measles, still others with diphtheria or whooping cough, or influenza. Because of that—that division of labor—the more MHC molecules we have, the longer we live in this hungry world. If our mothers and our fathers differ from one another among the genes inside the MHC, then we may get twice as many different molecules to stand between our selves and all the rest of the living. That's better. More inside the MHC is better.

Evolution understands that. Evolution made it possible for our mothers to smell these molecules on our father's shirts and pants, underarms and thighs. And evolution makes our mothers feel a little more like mating when our fathers' MHC genes are different from their own.

Those differences improve babies' chances for survival. But because our mothers are driven to seek men with different MHCs, it is nearly guaranteed that there will be genetic differ-

ences between each child's mother and father. Those genetic
differences appear on the fetus, making fetuses look a little like
their fathers. That is something mothers' immune systems cannot
ignore.

In the first few months of pregnancy, some of the fetus's cells
cross the placenta and move into the mother. This continues
throughout the pregnancy. Those cells, the baby's cells, carry the
products of the father's genes—the ones that the mother's nose
all but guaranteed would differ from the products of the mother's
own genes. When those cells arrive from the fetus, the mother's
immune system notices that the father's genes have found their
way inside of her, and her immune system responds, aggressively,
against the father. Eventually, the mother must speak of this to
her child. And when she does speak to her child, the antibodies
she sends against the father are mixed with antibodies that have
protected the mother from all the infectious and hurtful things
she has seen in this world. Antibodies against the parts of this
world that sought to hurt or kill our mothers—bacterial toxins,
parasites, murderous viruses, terrible trials our mothers have
faced. Messages of danger, and in the midst of them, words
about our fathers.

——

There is a point beyond which my relationship with my father
collapses. I realized it for the first time, though only dimly, the
morning we glared at one another across our bowls of gruel.
Many more times since, I have been reminded that I and my
opinions are valued only to a point—the point at which we dis-
agree. Beyond that lies a wasteland where my father is king and
I am his child. I have always resented that, that limited respect
my father bestows on me. For most of my life I have used that
resentment to defend my anger at him. But I know now that I
have placed limits on our relationship as well. I accept him as

he is only when it is easy. When he is not as I wish him to be, my acceptance shrivels. Then we stand on a barren plain, and upon that plain I am king, and he is a foolish interloper with no legitimate claim to my throne, or my mother.

His seems the more defensible position. After all, I am his child. And after twenty or thirty years of rescuing me from my screw-ups, it must be difficult, perhaps impossible, to transcend all of that and accept me as an equal, as responsible, as capable of taking charge of his life. My disdain, on the other hand, seems difficult to explain rationally.

"I don't like the place where we live," he says to me one morning over breakfast.

Immediately, I am angry with him. The place where my parents live is a retirement home I like very much and spent weeks preparing for him before his move to Fort Collins from Kanab. My mother is ill, he can't take care of her in a small town with no hospital. That's why I asked him to come. He notices my face tighten.

"I don't fit in with those people. They dress for dinner. They all have their own little cliques. They think they're better than me," and he pushes at his pancakes, refusing to look at me.

The air stinks of stale coffee and imitation maple syrup. We stink, too, of something much older.

He's wrong, I'm certain. He has worried much of his life about whether others imagined themselves better than he is. A waste of time. I can understand that he's just left nearly all he had in another state, and he's uncertain about everything just now—my mother, me, himself. But I'm still angry with him and his frailties, angry at him for growing old. Angry at him for questioning what I've done.

"I'm seriously thinking about moving."

"Dad, the place you have is great. There's an emergency pull cord in every room. You get your meals and they clean your apartment and change your linens every week," I reply, trying to hold my irritation below the detectable level.

"I don't like the food. I've made all of my own meals for the past month. Maybe I'll see if I can get my house back in Kanab," he mutters sullenly.

"Dad, you can't manage a house by yourself any longer."

Knowingly, I have just crossed over onto that barren plain. Now he's angry.

I never mean for us to arrive in this space. But too often we do.

I want to blame one of us for what is going on here. Mostly, I want to blame my father for his thoughtless words. But he is old.

And I? I am my mother's son.

———

Mothers speak to their unborn children with more than just antibodies. As soon as placental circulation is established, mothers send cells, too, to carry their words into their fetuses. It isn't clear what mothers' cells do inside of children, what stories those cells speak to children. But mothers' cells last for years inside of their sons and daughters, and those cells go nearly everywhere— lung, heart, bone marrow, liver, lymph nodes. Often, those cells are also found inside of children's thymuses. The thymus is the seat of the immunological soul, especially in developing fetuses and young children. When the mother's cells arrive in the thymus, the child is struggling for self-identity, striving—inside of the thymus—to separate self from not self. In the center of it all, there is suddenly the mother.

People with SLE (systemic lupus erythematosus) cannot reliably distinguish between self and not self. Because of that, these people's immune systems attack their selves as though self were the worst of enemies. People with SLE make antibodies against their own DNA, against their own genes and the proteins that hold their chromosomes together, antibodies against their own ribosomes where all proteins are normally made—antibodies against all the things that help make us who we are. Boys with SLE have more of their mothers' cells inside of them than do boys without the disease. Boys who've lost the ability to distinguish themselves from what surrounds them have more of their mothers' cells than do boys with clearer images of what is self and what is not.

Words of the mother's that live inside her children for years— words about self, words about not self, words about the father.

———

Maybe there is more to this than our harsh stares over a cheap breakfast. Maybe something else is pushing at us, urging us toward one another. Since life first coalesced in the warm seas of this world, we have been headed toward this point, this precarious moment perched over fried eggs and potato pancakes. For more than three billion years, things have been conspiring to bring us to this table. How can we ignore that, my father and I? How can we imagine that we control this moment?

An unimaginable time twisted by inconceivable forces. And all of that time, all of the conspiracy is still inside of us. It's in our genes and it's in the words we hurl at each other this morning. It salts the torn flesh of our relationship.

Time and biology have brought my father and me here, as surely as my father's Mercury. Brought us here, tossed us each a handful of nails, and left us.

———

Why would biology do a thing like that? What possible advantage, biological or otherwise, could derive from setting father and child at odds? Perhaps none. Perhaps all of this is just an accidental consequence of the way other things evolved—a simple side effect of an effort to enhance mate selection and mothers' protection of their offspring. Maybe we're just victims of an accidental intersection of biology and sociology. But humans aren't the only animals where fathers and offspring struggle to coexist. Other mammals, some fish, even some insects share with us our incompatibilities. When biological characteristics are shared by different species sometimes it is because those characteristics give each species a selective evolutionary advantage.

From a gene's point of view, it might be useful for a son or daughter to assist in steering a mother away from the children's father. The more males she mates with, the greater the variety of DNA in a woman's offspring. And the more males a woman mates with, the greater the odds that a mother's genes will find their way into future generations. At the same time, the father could easily see any children, especially children who love their mothers, as threats. Evolution can be a cruel god. But who knows?

———

In the end, whose hand is it that finally pushes the baby from the mother? There is the uterus itself, of course—the hormones that act on it, the fist the uterus forms. But there is also the father.

The way a mother's immune system responds to the presence of the father inside of her is the same way that her immune system would respond if the father's kidney or his liver were inside of her—the mother's immune system tries to reject the baby. The baby responds by trying to hide the parts of it that look like the father, take them from the trophoblast—where fetus and mother meet—and hide them. In place of those bits of the father, the fetus puts up other molecules that the mother cannot recognize as the father. And the fetus, just to be sure, secretes enzymes and other proteins that suppress the mother's immune system. The mother's self-portrait blurs, and the fetus prospers. But the mother's immune system, and its memory of the father, never stops trying to reject the child. In the end, that effort, that immunity, that rejection may be part of it, may be part of the mother's push, part of what shoves the child into a first winter night—into the light, into the vow.

——

Agreements were struck, oaths were sworn—things said and things shared between us and our mothers—things that no one understands. But things, too, that no one can ever take back. Before and after our births, my father and I learned things about our fathers that changed us forever. Just how it changed us is impossible to know. But there is an edge to my relationship with my father that is difficult for me to explain without invoking our biology. An edge we occasionally hone as sweetly as any barber about to scrape the stiff whiskers from a senator's grizzled chin.

——

Where three roads meet, Oedipus—imagined dead by his father—met a man with a wagonload of slaves bound for Thebes. The man was rude and aggressive and demanded that Oedipus move off the road and allow him to pass. Oedipus refused. The man struck Oedipus with a horsewhip. In his anger, Oedipus struck back with his staff and killed the man who had so offended him. Though Oedipus didn't know it yet, just as the Oracle at Delphi foretold to his mother Jocasta and his father Laius years before, Oedipus had slain the man who fathered him. Though crippled by his father's own attempt to murder him, though warned by the Oracle, Oedipus fulfilled his destiny—he murdered his father.

How is it that Oedipus, even though seemingly ignorant, moved so quickly to strike life from the man who had given it to him? And why is it that Oedipus then fell so easily into his mother's bed?

Sophocles never tells us what riddle the Sphinx spoke to Oedipus. But the audience knew, and they knew as well that those who could not solve the Sphinx's riddle were destroyed. Oedipus, though, knew the answer to the creature's question—it is man who walks on four legs as a child, two legs as an adult, and three legs when he is old.

Oedipus was destroyed anyway. He was destroyed by his own arrogance and ignorance; he was destroyed by the gods; he was destroyed by his fate; and maybe he was destroyed, too, by a bit of biology. Maybe it was just easier than he could possibly have imagined to kill his father and love his mother.

———

It is essential that we love our mothers. We—in some ways the most fragile of all mammals. Most of us wouldn't survive a day without our mothers' milk, their warmth, their fierce love of us. There is less need to love our fathers. The indifferent finger of evolution foresaw that and gave our mothers ways that all of those years of trial and error might speak to us through her— speak to us of our mothers and our fathers in very different sorts of words. Maybe that is why so many of us find ourselves at odds with our fathers. It seems that would be enough.

But there is more. There are the sins of the father as well. There are all the ways in which fathers have failed, though we never wished to. All the ways the wounds of the father are re-visited upon his children. Often, the harder we tried—so very like Oedipus—the worse we failed.

Our mothers smelled it—like the smell of the sea, rich and rank at once—the smell of our fathers. Inside of that smell our

mothers knew there were differences that would never be resolved, differences that must not be resolved. They married anyway. Then, they had to tell us about our fathers. They had to.
Evolution insisted on it. Inside the zygote, within the flame, in
the blood, in the milk, maybe in our births themselves the past
spoke to us, using our mothers' voice. And in words no one of
us could ignore, evolution told us how once our mothers were
girls who had no wish to be mothers and how our fathers had
changed all of that. From the beginning, each of us knew more
than we should have.

———

Oedipus knew the answer to the Sphinx's riddle, because Oedipus's father had tried to kill him, tried to kill him when he
learned what his son might do to his father. In that effort, his
father destroyed Oedipus's feet. Because of that, Oedipus walked
with a staff. And the staff gave Oedipus the answer to the
Sphinx's riddle. That saved him. But the staff had already be-
trayed Oedipus. The staff had killed his father.

His inability to forgive—either his father or himself for the
horrible things that had happened to them both—ruined Oedipus at last. Even his vow of poverty could not save Oedipus,
because pride had buried the vow.

We too, often because of things we controlled and things we
didn't, must consider the same questions that Oedipus faced, the
same challenges he failed. "How can I forgive my father for the
things he has done to me? Though he bolted my feet together
and left me to die, how will I forgive my father?"

———

The question, too, that might save us, the question that Oedipus
never asked. How will I ever forgive myself?

Whether we find answers to these questions or not, each of
us, like Oedipus, will be destroyed anyway. Nevertheless, we

must forgive our fathers and ourselves. It is not only a godly thing to do, it is our vow. Even our disappointments, even our anger, even our wounds themselves must be given away. After all, we promised. We promised that in return for even one instant of this sweet life, we would keep nothing of the world we were born into. Nothing.

Oedipus forgot that and paid with his eyes.

———

Today, almost thirty-five years after I looked on as he killed a mangled cat, I am again watching my father, watching my father watching my mother. My mother has soiled herself. It is probably my father's fault. She complained of constipation, and last night he gave her a laxative. It makes no difference. My mother is a mess, and she can no longer clean herself. Someone else must clean her. Today it is my father.

I am sitting in his living room, and though he doesn't know it, I can see into the bathroom where he is cleaning my mother. He is cleaning her as a mother cleans a child, as Michelangelo might have cleaned the last of the stone grit from the *Pietà,* cleaning her as old rivers scour stones—slowly, meticulously. He is cleaning her with warm cloths and lotion—across her hips, between her legs, down her thighs.

Though we have both cried over my mother's loss of mind, he is not crying now. He is simply cleaning her, carefully, washcloth in his fist, a small smile pulling at his lips.

Just now, I can only love him for that—for his smile, for the slow music in his old hands, for the way he is touching my mother.

My mother has honored her vow, she has returned everything ever given to her. My father and I aren't yet done. We are still giving things away every day. This morning it was innocence. But there is more we must rid ourselves of.

As I watch, I ask out loud, "Dad, can you ever forgive me for all of the things I've done to you? All of the times I hated you?"

"What was that?" he shouts to me from the bathroom, his hearing nearly gone now.

"Nothing," I say. Then after a moment, "Nothing that can't wait until you're finished."

Saved by Death

Eight years after I first tried to kill her, my grandmother finally died. She was sitting that morning watching an early snowstorm spit broken wheat across the empty face of southeastern Kansas. It was an hour or two before anyone noticed that she was no longer breathing. The doctor told my father that it was a stroke that took his mother from him. But I knew better.

———

Two days before, the doctor had told my father to go home. My father had just moved my grandmother into a nursing home. His guilt made him anxious to leave. The doctor said to my father: "Wayne, she may live like this for years. You should go home, be with your family." That was what my father wanted to hear. An hour later he left for our home in Bountiful, Utah—a thousand miles away. But the doctor was wrong.

The day my father arrived home, the doctor called. My father was too late now, but in the dark, he threw his things back into our 1960 white-and-imitation-wood Ford station wagon and drove off. He would do what he could. It wouldn't be nearly enough.

When my father's anger finally guttered, he talked to the rest of us some, talked about death. He told us that it was a stroke that killed his mother, and he told us that death was a part of life, something every living thing would face one day. We all felt better. But my father was wrong. People don't die because they live. People live only because they die.

————

Just south of the Wasatch Mountains, near where I grew up in northern Utah, lives one of the world's most massive beings. His name is Pando. Pando occupies more than 106 acres of land and he weighs over 13 million pounds. That's about 722,000 times heavier than either of the miniature schnauzers who put up with my wife and me, roughly 50,000 times heavier than Andre the Giant, 30 times heavier than blue whales (the largest of mammals), and 3 times heavier than the largest sequoias. By anyone's standards, Pando is huge.

But large as Pando is, his weight isn't the most remarkable thing about him. The most remarkable thing about Pando is his age. It is likely that Pando is more than a million years old—more than 1,000,000 years old. Pando has probably stood in what is now Utah since before there were men and women in North America, has likely seen the constellations change shape, felt the shift of the Earth's magnetic poles, and shivered as the interglacial ice advanced and retreated.

Older and larger than we can imagine. But nearly every one of us knows one of Pando's relatives. Pando is a clone of quaking aspen trees, a male clone. He was named and first described by Michael C. Grant, a biologist at the University of Colorado, Boulder. *Pando* is a Greek word meaning "to spread." Based on the resemblance of modern quaking aspens to fossilized ones, Burton Barnes of the University of Michigan estimates that some aspen clones are more than a million years old, maybe a lot more. In effect, aspens are immortal, dying only from destruction of their environments, fire, or disease.

In spite of Pando's size and age and leaves, he is really not very different from you and me. Like us, quaking aspens are complex multicellular organisms. They breathe air and carbon dioxide. Aspens' genes are encoded in DNA. They have circulatory systems; they photosynthesize; they respire; they consume carbohydrates, proteins, and lipids; they produce hormones; their cells have mitochondria and ribosomes and Golgi complexes and nuclei—just as our cells do. But, unlike us, quaking aspens reproduce like strawberries through a process botanists call vegetative reproduction. An individual aspen stem (what we call an aspen tree) puts out roots that, at some distance from the first stem, will themselves root and put up another stem. A *grove* of aspen trees is, in fact, one individual—just as a collection of cells combines to create one human body, one human person. But, in spite of all these similarities, there are two important differences between human beings and quaking aspens: humans have no leaves, and aspens are immortal.

The first of these is relatively unimportant, unless you need to photosynthesize. The latter, of course, changes everything.

———

In the hallway of the house where I grew up, above an oaken phone stand, my mother kept a wooden cross. With brads the length of beetles' teeth, a wooden Christ was nailed to the face of that wooden cross. A palm frond was twisted behind his head, a crown of tiny thorns pierced his tiny scalp, a trickle of wooden blood rose from the wound between his wooden ribs, and when you pushed on Christ's feet, the face of the cross slid back to reveal a false bottom—a secret space behind the body of Christ. As often as I was allowed, and more often when I wasn't, I took it down from the wall and pushed Christ from the cross. Inside, cradled in red velvet, were two candles of pure beeswax, a cruet of holy water, and a vial of oil. I loved that cross and the secrets it held, I loved to open it, close it, roll it in my hands.

In all the time that I lived in that house, I saw the cross truly

used only once. It was a September night when my father's
mother was near dead. I was fourteen. The parish priest came
to our house. My father met him at the door. Both looked very
serious. They spoke in hushed tones, and I could make out none
of what was said. My mother sat by herself in the kitchen.

At last my father nodded toward the wooden crucifix, and the
priest took down the cross and pulled from it the body of Christ.
I was rapt. The priest set Christ carefully aside, then lit the two
candles, and placed one at each side of my grandmother's bed.
He muttered a few words of Latin, mixed the oil and water in
the palm of his hand, and rubbed the mix into my grandmother's
forehead. He continued muttering in Latin, as he made the sign
of the cross over my grandmother's head and chest. I prayed along
until my father noticed me there beside his mother's bed. Then
he pushed me from the room, into the hallway.

I stood, embarrassed by the rejection, staring angrily at the
beige paper that lined our hall. But I refused to move further.
And while the priest mumbled and the house began to stink of
candle wax, I saw on the wall, where the cross itself had hung,
a ghost cross—an image in the faded paint. I had seen it before,
this phantom cross carved by sunlight into the wall above the
phone. But tonight it looked different to me, somehow more
important. I thought for a moment, and then it came to me. The
image of the cross was like those I'd seen in school, the ones
etched into the bomb-bleached walls of Hiroshima and Nagasaki.
Except those ghosts were shaped like people and houses, auto-
mobiles and dogs and cats. This one looked like a wooden cross.
At first, that pleased me, thinking of all those bombs, those fire-
balls, all of that destruction. It fit my mood. But as the after-
images clarified behind my eyeballs, I remembered that earlier
this morning, my grandmother and I had fought over when I was
to mow the lawn. She had forced me into it. And while I mowed,

I cursed her. I wished her dead, sincerely. Now I was frightened by what I'd done. Ashamed and frightened that it was my wish that was about to be granted.

I wanted to tell my father, but he only scowled at me when I tried to reenter the bedroom where he and the priest prayed over my grandmother. I cringed and slunk back into the hall. And while I stood there chewing my cheek, I alone knew that when my grandmother died here tonight, it would be because of me.

———

Eight years later, the year my grandmother finally died, somewhere around two million people died in the United States. Of those, about two-thirds were, like my grandmother, over the age of sixty-five. Perhaps pathologists understand why those people under sixty-five died in 1969. But in spite of what the doctor told my father, in spite of the lists compiled by the Centers for Disease Control and Prevention, in spite of the curse I'd conjured for my grandmother, none of them knows why the million and a half or so people over sixty-five, including my grandmother, died in that year.

Admittedly, we know *how* most of those people died. Keeping track of these things is the job of the people at the Centers for Disease Control and Prevention. Some people died of heart disease, some died of cancer, some died of strokes, and so on. We know a lot about how these people died. It's just that we don't know why.

———

But everyone thinks he or she knows why people over the age of sixty-five die. They die because they are old. Old people die, everyone knows that. And everyone knows that a million and half or so people died in 1969 because they were old. It's as simple as that.

Even though the epidemiologists at the Centers for Disease Control and Prevention do not list "old age" as a cause of death— even for people over the age of sixty-five—all of us will die, and most of us will die of "old age." That is the nature of human beings; we die. And we've been dying now for the better part of three million years. Death is an inescapable consequence of life. The old die, we all know this.

Of course, none of us learned it from a dead person.

———

Bacteria are the most successful creatures living on this planet, no matter how you measure success—species diversity, mass, distribution, longevity, sheer numbers, adaptability. You name it. By every standard, except assets, bacteria are the most successful things that have ever lived on this Earth. More than 99 percent of all living things are bacteria. And worldwide, there are more than twenty trillion tons of bacteria. This is and always has been, as Stephen J. Gould said, the Age of Bacteria.

And like Pando, the most common of all living things don't die. Bacteria don't die from "natural" causes, don't die from old age, either. Bacteria simply grow and divide, grow and divide, grow and divide. The oldest and most successful living things on the face of this planet don't die.

Bacteria are prokaryotes, which means they lack a well-defined cell nucleus. The eukaryotes, organisms whose cells have a nuclear membrane and easily recognizable cell nuclei, form a much smaller kingdom than either of the bacterial kingdoms, the archaea and the eubacteria. But the eukaryotes are considerably more interesting to most of us, because humans are eukaryotes. So are funguses, plants, and all the animals. Some eukaryotes, including funguses and some plants, don't die either, not from old age anyway.

Life began without death. But somewhere between bacteria and humans, or perhaps even later—somewhere between funguses and humans—living things learned to die. And ever since then, death has been our way of life.

———

Genesis 3:19: "By the sweat of your brow you will eat your food until you return to the ground, since from it you were taken; for dust thou art and to dust you will return." And Genesis 3:22: "And the Lord God said, 'The man has now become like one of us, knowing good and evil. He must not be allowed to reach out his hand and take from the tree of life and eat, and live forever.'"

It was a serpent that led Eve, then Adam, to the Tree of the Knowledge of Good and Evil and induced them to eat. But it was a god who condemned them to death.

In Orphic myth, the Titans ate Dionysus, the son of Zeus and Persephone, and in his anger Zeus destroyed the guilty with a thunderbolt. From the Titans' ashes came men and women. And because they rose from ash, in spite of their divine origin, it was the fate of men and women to return to ash.

The first Nigerian kings were gods. The ancestors of Diné, Lakota, and Algonquin men and women were gods as well. The first aboriginal peoples of Australia were immortal, too, and given a single rule—don't eat the flesh of animals. But when drought struck and the plants failed, there were some who fed upon the animals. It was then that death arose in the form of Yowee, the death spirit.

Creation myth after creation myth speaks of immortal ancestors and mortal men and women—divine origins and a condemnation to mortality.

———

Science, too, speaks of immortal ancestors—bacteria—undying ancestors from which we were born. But what of the condemnation? What of the origins of the dead?

If you take skin from a human being and place it in a dish containing, in essence, artificial serum, the living cells in the skin will grow and multiply. For a while. But after each cell has divided maybe ten times, maybe twenty, all the cells will abruptly die. The same is true for liver cells, pancreas cells, intestinal cells, muscle cells—all normal (noncancerous) human cells. Human cells live for a few weeks or months and then all of them abruptly die. This occurs independently of time and metabolic issues and seems to depend only on the number of divisions the cells have undergone. That is, there seems to be a mitotic "clock" inside of normal cells that triggers their senescence and death after a fixed number of divisions. That clock depends on the disappearance of an enzyme called telomerase. In very young cells, telomerase protects the ends of chromosomes (called telomeres) from degradation by other cellular enzymes. But evolution has seen to it that, as cells mature, telomerase disappears. Once that happens, each time the cell divides, a little of the DNA is clipped from the telomeres of the chromosomes by cellular enzymes (also provided by evolution). When enough DNA is gone from the chromosomes, the cells stop dividing and die. The result is mortality. What we did to deserve it no one remembers.

But we must have done something, because bacterial cells—our ancestors—don't die. Bacterial cells grow forever in the laboratory and outside of it. Bacteria can be killed by lots of things, but not by time alone.

Cell biologists call this characteristic of human cells "cellular senescence." They call this characteristic of bacterial cells "immortality."

Inside of humans, a similar thing happens. Cells grow and

divide for a while and then they die. Skin cells die, and they die at exactly the same rate that new skin cells come into being. If they didn't, very soon we would have enough skin for two people, and then enough for four, and then enough for eight, and so on. Same for liver, same for kidney, same for spleen, same for white blood cells like lymphocytes and neutrophils, and the same for most all of our other tissues and cells. Human cells grow for a while and then they die.

But it isn't age that kills these cells. Human cells don't die because they grow old. Some human cells, like nerves and muscles, live for a human lifetime. So for those cells that die, death is not simply a consequence of life. Human cells die because death is in their genes.

Lymphocytes are the cells that drive our immune systems. Lymphocytes make antibodies, destroy viruses, fuel inflammations, respond to vaccines. Lymphocytes are what make human life possible in an infectious world. None of us could live for more than a few days without lymphocytes. But lymphocytes are born dying.

Human lymphocytes begin life in the bone marrow, the red stuff that fills our femurs and ribs and a few other bones. And at the instant of a lymphocyte's birth, a clock is set. Without outside help, that clock will soon strike midnight, and that lymphocyte will die—destroyed by the artillery hidden inside its own DNA. For lymphocytes, life isn't about living, it's about how to avoid dying.

Lymphocytes are prewired to self-destruct, because they are a threat to all the other cells in our body. Shortly after they are born, lymphocytes become deadly, fully capable of killing other cells in our bodies. They must be, because human babies need to be protected from all that would harm them—measles, staph, strep. But lymphocytes are blind, they have no way of distinguishing between friend and foe. No means for deciding what is baby and what is not. Newborn lymphocytes are as likely to rip holes in our livers as they are to protect us from influenza. This is a very dangerous situation.

In spite of the danger, we cannot live without lymphocytes. It is only because of these cells that individual men and women exist. Mice born without some types of lymphocytes will accept skin grafts from chickens and grow their own feathers; without lymphocytes mice cannot distinguish between themselves and chickens. But for our own protection, many lymphocytes must die.

A lymphocyte is allowed to live only if it can prove itself trustworthy. From the bone marrow, lymphocytes have two choices—they can move directly into the blood and become B lymphocytes, or they can migrate to the thymus, an organ that sits just above the heart in a human and is about the size of child's fist, where they become T (for thymus) lymphocytes. In the thymus, lymphocytes face the first jury of their peers.

The thymus's first task is to determine if newly arrived T lymphocytes are functional. The thymus's second task is to eliminate from the functional cells those with a taste for human flesh.

So when a lymphocyte arrives in the thymus, the thymus first checks to see if that lymphocyte is a killer. Lymphocytes that cannot kill are of no use.

If the lymphocyte shows that it is capable of killing, the self-destruct program inside the cell temporarily stops. But if the T lymphocyte hesitates and fails the test, then it receives no help from the thymus. Then, the self-destruct program running inside that cell proceeds, and abruptly the lymphocyte's DNA falls apart, its membrane bubbles off, and the lymphocyte dies an ugly death.

The lymphocytes that pass this first test must immediately pass a second. The thymus knows that the remaining lymphocytes are natural-born killers, blind natural-born killers. The thymus knows, too, that any one lymphocyte will recognize and react to one and only one thing for all of its life. That means that if the thymus can find out whether a lymphocyte is looking for polio or human blood, that lymphocyte will always look for polio or human blood, never both. The thymus finds out by forcing each lymphocyte's hand.

While the thymus watches carefully, each lymphocyte is shown the cells of the thymus itself (immunological self), like murderers being shown their victims. If any lymphocyte so much as twitches, the thymus starts the self-destruct program all over again. So that even as these cells try to destroy the thymus, they destroy themselves.

On the other hand, lymphocytes that seem only mildly and pleasantly interested in the thymus, act as if the thymus is some distant relative, those lymphocytes are allowed to live, for a while, and pass out of the thymus into the blood.

Even after all that, infection itself or the absence of infection may kill T cells that survived the thymus. But none of these T cells die by accident or because of age. T lymphocytes die because of a sophisticated biochemical program evolved over millions of years. Evolved and refined by natural selection because it works, and because it protects us from a fate worse than death. It's in our genes, and if it was any other way, we might well die at the hands of our own lymphocytes. Death is our ally, in this case our only ally.

This self-destruct program, the one used by lymphocytes, is called apoptosis, and apoptosis kills more than lymphocytes. As we develop inside our mothers, apoptosis kills the skin between our fingers and toes so we aren't born with webbed hands and feet. Apoptosis kills skin cells and intestinal epithelium, kidney cells and lung cells, macrophages, neutrophils, and megakaryocytes. It kills them inside of our bodies and out. Not because any of these cells have grown old, but because the unrestricted growth of these cells, their immortality, is a threat to our lives. For the good of the whole, these cells must die.

The importance of this preprogrammed death is made grimly obvious in human cancers. Cells whose apoptotic programs have been turned off by a genetic accident become tumor cells— malignant or benign, lethal. Cells without self-destruct programs grow into uncontrollable masses, they steal food from healthy cells, they crimp arteries, they crush brains, they crack bones, they destroy livers. Cells that don't kill themselves kill us. People that didn't die would kill us just as surely.

We are large omnivores. Our eating and living habits have a considerable impact on the world around us and on each other. In a relatively short period of time, immortal humans would outstrip the capacity of their world to feed them. We live only because we die. Our lives depend on death—the deaths of our cells, the deaths of our bodies.

Immortality is toxic, except for bacteria, funguses, and, of course, Pando.

———

Quaking aspens, like Pando, are the most widely distributed trees in North America. Groves of these trees stretch from Newfoundland to Maryland, from Alaska to Washington, along the Appalachians south to Georgia, and along the Rockies into Mexico, covering millions of acres. Aspens thrive in damaged habitats—after mud and snow slides, and especially after fires. In fact, without fire, even the greatest of the quaking aspens is rendered mortal. Fire clears new ground and slows the invasion of conifers—pines and firs—that cut the sun. Aspens can't tolerate life without sun.

The name "quaking aspens" comes from the way the trees' leaves tremble in even slight breezes. Four hundred years ago in French Canada, woodsmen wrote that aspens quake because Christ died nailed to a cross of aspen wood. They said it was the memory of a god's death and the ghost of his immortal spirit that caused the trees to tremble. They said it was the cross itself, and the ghost beneath it, that makes the aspens quake.

——

My mother has had two heart attacks. My mother is old, eighty-two. The doctor told my father that her next attack will be her last. "And there *will* be a next one," the doctor said. "She is old, you know." What he meant was: "Old people die, and your wife is very nearly old enough. What you have taken for granted for almost sixty years is about to leave you. You won't know what to do when that happens. Believe me."

About the first part, the doctor is wrong. My mother is not dying because she is old. My mother is dying because humans rose from the ashes of a god. Dying because, just as the thymus cannot grant immortality to lymphocytes that are unable to distinguish between self and not self, the gods could not risk immortal men and women who had eaten from the Tree of the Knowledge of Good and Evil and believed they understood good and evil. She is dying because evolution in its nearly infinite wisdom has granted each of us the gift of death. Life is toxic. Immortality is for the few.

——

It was over thirty years ago that the priest pulled that wooden crucifix from our hallway wall and ministered to my grandmother. Over thirty years ago that my grandmother survived what I imagined was a fatal curse. Three decades—the blink of an eye for Pando, a sizable portion of a lifetime for me. I realize now, after most of a lifetime, that my curse had little effect on my grandmother's life. I would be relieved by that—the fact that I played no role in my grandmother's death—except that now I know for certain that her genes are killing me.

The Metamorphosis

One evening in the summer of 1964, Harold Patterson walked into his basement wood shop, turned on his table saw (carbon steel, cross cut blade), waited a moment for the saw to come to full speed, and sawed off his hands. Sawed them clean off—skin, ulna, radius, skin—clean off. Harold's hands fell beside the rip fence. For an instant, I'm sure Harold thought about turning off the saw. But even as the thought formed, he must have realized that was no longer an option. So Harold walked back upstairs and showed the stumps of his wrists to his wife Elaine. Because she could think of nothing else to do, Elaine ran to get the neighbors—most of a mile away. When she returned, Harold was dead. When the police arrived an hour later, the saw was still running in the basement. One of the policemen walked downstairs, turned off the saw, gathered up Harold's hands from where they lay, and dropped them into a clear plastic bag.

After that, the neighbors took Elaine home with them. The police locked up Harold's house and took what a few hours earlier had been Harold to the morgue—body in back, hands up front. Then the policemen, too, went home, held their wives a little more tightly than usual, and slept fitful sleeps. It was a terrible thing.

———

Harold was my uncle, my mother's older brother—one of eleven children. Among my mother's brothers, Harold was my father's favorite. Harold was an offensive tackle on the right side of the Carbondale Junior College offensive line. My father was right end. My father told me that when he thinks of Harold, he thinks of high school football.

Among my mother's brothers, Harold was my favorite, too. We played catch together. When I think of Harold, I think of a man sawing off his hands.

Why would anyone do a thing like that?

Maybe it was something he touched—something so terrible (or so wonderful) that Harold never wanted to touch anything again. Or maybe he was just fatally tired of holding on to things, picking things up and putting them down again, moving things from one place to another for no good reason.

Maybe it was none of that. Maybe Harold never meant to saw his hands clean off, only to nick his wrists and draw the deep red blood to show to Elaine and just see what that woman would do. Maybe that was what he meant to do, but at the last instant, he was pushed, pushed by that same someone who pushes at each of us as we stand a little too close to an open window on the sixteenth floor. Someone who insists we can fly. Someone who whispers that holding on to a carbide-tipped steel blade would be no different from grabbing up handfuls of chiffon. Someone dead anxious to end this life and get on with the next.

————

It is five years ago. I am sitting on the floor of our kitchen leaning against a leg of the hard ash table there. The room smells of rubbing alcohol and this morning's eggs. I am holding our dog, a miniature schnauzer named Nell. The dog is dead. And I know already that for as long as I live, some part of me will always be sitting here holding our Nell. A moment ago, she was alive. No, that isn't quite right. A year ago she was alive, but for the past twelve months, pieces of her have been falling off. A disc in her back prolapsed, her pancreas shut down, her eyes and her ears quit on her, and a blood clot or two lodged in the capillaries of her brain. A moment ago we scooped up what was left of Nell and stopped the cruel fingers of this beautiful world from picking at her any longer. A moment ago, we filled what was left of Nell with pentobarbital. And though I know that what we did was the very best of the things we might have done for Nell, just now I wish we hadn't.

Gina and I spend the rest of the afternoon telling one another stories about Nell. This helps with our grief, but it won't bring Nell back.

Gina reminds me of the day our Nellie ate an entire garage. Well, almost. One afternoon, we left Nell alone in the garage of Gina's condominium in Imperial Beach, near Tijuana—just one afternoon. When we got back, Nell had eaten into the exposed dry wall, as deep as her teeth would pierce, as high as her mouth could reach. Nearly an entire garage. Her beard was filled with gypsum and scraps of paper. She was happy to see us.

Another time, Nell ate *The Ends of Power* by H. R. Haldeman and *Annapurna* by Maurice Herzog. She also ate Waylon Jennings and Jessie Coulter's *Outlaws* album, a cannister of 35-mm film, my son Patrick's baseball autographed by all of the 1976 San Diego Padres (including Cy Young Award–winner Randy Jones), my journal, a copy of *On the Road* by Jack Kerouac, an overstuffed reading chair, *Science* and *Bon Appétit* magazines, *The Rise and Fall of the Third Reich*, *Moby Dick*, and bits of children when she could. Small and gray with fierce brown eyes, our Nell was eclectic. She made us laugh. Even in her death, she made us laugh.

There were others after that—our friend Melanie, our neighbor Phil, Gina's mom, Lila. I miss them all, all of the dead. That's selfish, I suppose, childish. But seeing that, seeing that my anger is no more than the anger of a child whose favorite things have been hidden from him, makes little difference.

———

One day five years after Nell's death, Gina tells me she still sees Nell, sometimes playing in our yard. "At first," she says, "I always think it's one of the other two dogs—Tamale or Crow. But when I look, both of them are asleep in the chair in the living room. So I know it's Nell. But then, when I look back, she's gone."

I believe her. Gina is a clear-headed woman. If Gina tells me so, then I believe Nell still plays in our backyard. And whether I see her or not, I would give anything I own if someone would allow Gina to run with Nellie one more time, to hold her, to speak to her. (Nellie loved me, but she and Gina were part of one living thing.) Anything I own. Just one more time.

And then whatever was left over, I would give to spend one more afternoon with Harold in the wooden swing beneath his weeping willow, one more hot afternoon breathing in the moist midwestern air. We wouldn't even have to speak. It would be enough to just sit there.

Trouble is, nothing I own is worth near enough to raise the gods from wherever our forgetfulness buried them, raise them up and wring a promise out of them. Nothing I own can bring Harold and Nellie back from the grave. Nothing.

Or nearly nothing.

———

On the average human being there are a lot of bacteria—as I've said, somewhere around 10^{14}. Bacteria are little bits of life—a membrane, a chromosome, a few ribosomes (for making proteins), and a bagful of enzymes (the proteins they make). Bacteria live pretty much everywhere, including on us. 10^{14} bacteria is more bacteria than there are stars in the sky. That's more bacteria than there have been seconds since *Homo sapiens* appeared on this planet. That's more bacteria than there are grains of sand in a square kilometer of beach. That's more bacteria than any human can imagine.

Even a large human body contains only about 10^{13} to 10^{14} human cells. Myocytes, hepatocytes, lymphocytes, adipocytes, megakaryocytes, basophils, eosinophils, neutrophils, neurons, glial cells, red cells, osteoblasts, osteoclasts, monocytes, basal cells, squamous cells, columnar epithelium, sperm, ova, eyeballs, bone, bowel, teeth, hair, and nails. That's a lot of cells. But at best, it is only 10 to 50 percent of the cells that occupies the space we call us. One-half to 90 percent of the cells we call us don't, in fact, belong to us at all, don't belong to anyone.

Staphylococcus epidermidis and *aureus* on our feet and hands, on our groins and scalp. *Streptococcus viridans, pyogenes, mitis,* and *pneumoniae* in our mouths and throats. *Lactobacilli, Streptococci,* enterobacteria, *Bacteroides,* fusobacteria, *Escherichia fecalis, Escherichia coli, Klebsiella, Pseudomonas, Clostridium,* and bifidobacteria in our stomachs and intestines. 100,000,000,000,000. And on top of all that, there are funguses like *Candida,* like athlete's foot, like ringworm, like *Aspergillus.* And often, there's a protozoan or two like *Trichomonas* or *Giardia,* and maybe a worm or two hundred.

All of those cells are feeding on us in one way or another—acquiring our amino acids, carbohydrates, nucleic acids, fats, carbon, hydrogen, oxygen. Some of them are feeding us in return. Who can say what is the smallest piece of me that still qualifies as me? A lifetime of sharing, though—even a short lifetime of sharing—and it's obvious that much of what I think of as me has spread between me and 10^{14} other beings.

But that isn't the real issue either. Nor is the fact that our individual deaths don't alter in any mathematically significant way the number of living things that occupies the space we call us.

The real issue is that all of this is beyond imagining, well beyond the arbitrary cap this universe placed on our ability to envision the very large and the very small.

Suddenly, it's *us*. The cells that I call "me" are mostly bacteria and funguses and parasites. Inside my own body, I am the ultimate ethnic minority. I'm surrounded, inside and out. I'm never anywhere near alone. I am a community, an entire ecosystem. I am a world apart, except for the "apart" part. A collective. And when I die, when the one I call me is gone, there will still be more than 100,000,000,000,000,000 living things lying where I fell and all of them will be filled up with pieces of me.

Even cremation probably doesn't destroy much more than 99 percent of those bacteria. That still leaves 10^{12} living pieces of what was recently called me swirling in the plume above the furnace. Dancing in the light.

———

Harold was thin and often brown as the summer dirt. His hands were warm and long-fingered, his hair a dark shock that the wind shook out like horsehair above his high forehead. That much of Harold I still have. But I've lost the color of Harold's eyes, the music of his voice, the spread of his feet and shoulders, his broad brown back, and who knows what else?

No other human act is as powerful as forgetting. Death pales by comparison to forgetting. Forgetting has buried unknowable thousands of us. Forgetting has emptied whole vistas. It has crushed civilizations. It has erased entire species of living things. It has emptied languages. It has even destroyed the gods themselves. Remembering is a fragile instant. Forgetting is eternity.

Like the Oregon Trail, each of our paths is littered with stuff we thought we could get by without—an iron stove, a Hoosier cabinet, a keg of nails, an oaken chest, Grandma's iron tub. It was all just too heavy and the mules hadn't eaten for near a week. So we threw it out. And then there was the stuff that simply fell off somewhere along the way—lanterns and shoes, augers and tack, candles and tubs of lard, our single cast-iron skillet, and that odd child with the piercing blue eyes. Sometimes we hated that, losing all of that stuff. And sometimes we simply forgot it. But we made it—lighter for certain, considerably lighter—and that's the important part.

It's good that we made it. The problem, of course, is that now we can't even recall what it was that we tossed overboard, let alone the junk that jumped ship on its own as we rolled across the plains or over the mountains, stuff that we dropped crossing the turgid rivers or things that were flung from the wagon when Uncle Jacob shattered a wheel among the rocks.

Who knows how many died or disappeared along the way, how many places known only to us vaporized while no one else was watching? How many horses, how many cows were lost? Whose touch have we forgotten? Whose breast or lips, whose breath on our skin? Who knows?

No one, obviously. No one knows.

————

And then there's the truly indestructible stuff: like carbon, hydrogen, and oxygen. Stuff like nitrogen, sulfur, phosphorus, and potassium. Others, too, that I've forgotten. The stuff of proteins and nucleic acids, complex carbohydrates and saturated fats. Stuff entrusted to us by the stars. And stuff that somehow makes all of us the fanciful things we are. Every bit of this Earth that was us we leave behind, too.

Every rusted atom of iron ever given to any human came to him or her used, and every one of those atoms is still here, somewhere. Only forgetting is final.

It's hot. Clouds, like black bears at the dump, are snuffling and scratching at the low hills to the south. On days like this, the wind fills up with water and electricity and tastes like new tin or old dirt. I am twelve years old, and it is the Fourth of July. Even better, earlier today my brother Michael leveled one of my Uncle Harold's elm trees with a power mower. Michael didn't mean to do that, but while driving Harold's tractor-mower, Michael's mind moved off to the long straight flats of Bonneville southwest of Salt Lake City for a run at Craig Breedlove, the "Spirit of America," and the world land-speed record. With his mind in the desert, Michael failed to notice the sapling that loomed ahead. As he took the last of the straightaway and raced for the timing pins, he sliced through that little elm like Genghis Khan might have leveled an offending child, and then the mower chewed the fallen into pieces no larger than the average human thumb.

My father was furious. Harold laughed and then stood between Michael and my father.

Dinner intervened, and my brother was saved. But in those few moments, my stock had risen considerably. My older brother was for the moment dethroned. And for an instant, it appeared that I was next in line for the crown.

Dinner was barbecued hamburgers, and before we were done that day I'd eaten five. Harold seemed pleased with such an overt testimony to his culinary prowess. My father seemed pissed. But only *seemed* pissed compared with how he was about to be *genuinely* pissed.

The clouds overhead are thickening now and have begun to acquire that evil greenish skin I've only seen on top of old soup and in hail clouds over the Great Plains. The wind is getting dead serious about clearing all the land from here to Illinois. I am moving slowly toward the last of the coals in Harold's red brick barbecue—I have a plan. I look carefully to be sure no one is watching me, and then I fling a string of a hundred Black Cat firecrackers into the barbecue.

I move back. For an instant, everything is dead quiet. Even the wind stops. The Earth rotates a few degrees east, and the solar system spins a few more miles round the black hole at the center of the Milky Way—but no one besides me notices. It is as though each of the pieces of this world has finished with the present moment but isn't quite certain it's wise to move on to the next.

Then there is sound, a sound like none other I've heard— before or since. There is a certain Chinese quality to this sound because of the fireworks popping in the barbecue. There is also a certain Caribbean character to this sound because of the gale-force winds and the rain that has begun to fall. Then there is an underlayment of the New Jersey waterfront because of my father's cursing. Finally, there is a certain southern charm to this sound because of the women screaming in shrill notes behind me. Above it all is the melody of my uncle Harold's laughter, crisp as apples on October mornings.

The visuals aren't bad either. The barbecue is erupting like Vesuvius. Sparks and chunks of burning charcoal are spinning through the air and landing everywhere. All of the Earth that is not at this moment sparks is bone-dry grass the color of kindling. People are dodging flying embers, the wind roars, my father curses, the clouds darken, hail begins to fall, steadily pelting everyone. Harold laughs out loud, laughs as the angel Gabriel might laugh just after playing his solo piece in front of God and everyone, just after playing it perfectly.

Sic transit gloria mundi. Oddly enough nothing catches fire.

If he could have, my father would have ended it all there. I am alive today only because of my Uncle Harold. The dead man with no hands.

———

Our water remains after us, too. Water that is us as much as it is anything. The water we drink, the tears we shed, the sweat we offer, the saliva that falls from our mouths onto our lovers' skins, the water that moistens us, the water we share. That, too, will stay for those who remain behind.

Though water lasts nearly forever, most of the water on this world came to us from someplace else. Only a tiny fraction of it was created here, and only a tiny fraction of it will be destroyed here. The rest came to us from the stars—hydrogen from the plasmic fires of the big bang itself and oxygen hammered from hydrogen on stellar anvils. The water in your cup was once my grandfather's tears. The water poaching my salmon was once your mother's milk. Our water, water that was once ferns and pterodactyls, gray-green moss and blue lakes inside of stone-laden glens. Our past and our future. It is what we are. Seventy-five percent of what we call us is water, 75 percent of what isn't bacteria and funguses and parasites is nothing but water.

And when we die, our water stays behind. In the plume above the furnace, all of that is there as well.

Every breath we've ever drawn, too, will still be around when we are gone. Every one. The air we used to speak of love, the prayers we spoke aloud, each poem we shared, every lie we told, each truth and each hope we ever spoke of—all of those, too, we leave. Everything inhaled is exhaled. Everything gathered, released. None of it will go when we do. Remember, we agreed to that before we were even born.

———

So in spite of the length of our trip and the things we had to leave behind, there are a lot of things we carried to the end. The tintype of the young woman with the black hair, the woman no one any longer recognizes; the horses; the old writing table with Ted's initials carved into it, and all the rest. The stuff we couldn't have thrown away even if we had wanted to. Whether we wish to or not, that stuff—the very elements that carried our thoughts and wishes, the stuff that spoke our words, the things that lifted us up and laid us down—we will leave for others when we die. All of it—for anyone who might care to notice. That is a great deal to leave behind.

So much so, in fact, that naming exactly what the difference is between a living person and a dead person is very difficult. The medical profession speaks of two types of death—cerebral death and brain death. Cerebral death is the loss of all higher brain functions but the retention of physical homeostasis—death of the cerebral hemispheres (the outermost part of the brain that is associated with all higher functions) but continued functioning of the brain stem and cerebellum (associated with basic life functions). People in this state can breathe on their own, maintain their own heart rhythms, etc., but they are incapable of all those things we normally associate with higher brain function—speech, recognition, responses to many external stimuli, poetry, prayer. Brain death, on the other hand, is accompanied by the loss of all mental and physical function. Brain-dead people must be supported with a ventilator because they have lost the ability to breathe on their own and their brain no longer exhibits any electrical activity. Those bits of science and technology—spawned by our own fear of death—have changed the way we die.

Now there are rules about how to tell a dead person from a live person. There's cerebral death (where the human seems to have vanished), there's brain death (where the animal that rested beneath that human has evaporated). And there's dead death, obvious to nearly all of us—where all that remains is nearly everything that was there to begin with.

What's gone that was there a moment ago? Only the dead know for sure. But you can show that certain processes cease. Most important among these seems to be the process that we call oxidative phosphorylation—the use of oxygen to manufacture energy from fats and sugars inside of our cells. Most all of the signs of life—locomotion, sensation, response, respiration, and so on—depend on energy produced through oxidative phosphorylation inside of the mitochondria inside each of our cells. Oxidative phosphorylation, in turn, is dependent on normal breathing, adequate nutrition, appropriate signals from the brain or other elements of the nervous system, and circulation of blood. When any one of these fails, oxidative phosphorylation ceases, we grow cold, and we die. We don't die directly from the lack of oxidative phosphorylation; we die because, without it, our hearts cease to beat, blood doesn't circulate, and hypoxia (decreased oxygen supply) quickly kills the cells in our brains that control all of our other vital functions. Although it may seem a little circular, that's pretty much how it is. So the process that we call death is marked by the end of certain processes we associate with life.

Death is also marked by the initiation of certain processes. The body cools because of the cessation of oxidative phosphorylation. Then the muscles stiffen, because of a chemical reaction. Then the muscles relax as that chemical reaction ceases and autodigestion begins. Cells burst, and more digestion occurs. Finally, and perhaps most noticeably, the reactions that we associate with decay begin. These are the reactions that we can smell, and they result from enzymatic digestion of human tissues, digestion mostly by bacteria, and these reactions result ultimately in our gradual transformation into even more bacteria, funguses, parasites, sky, earth, and rivers.

Some processes end, others begin.

———

What, then, comes between the living and the dead? It seems to me that it is nothing more than *me* that stands between me and Harold. It is my own forgetfulness that is killing people, that and an imagination that won't hold a number like 10^{14}, that can't reach its arms all the way around the past and hold it here in the moment. Me.

———

Gina grieves more purposefully than I do. She understands the process and each of its pieces. She is skilled at it. I simply resent it. I resent the cracks that grief opens in us, and I'm afraid of the stories that we tell to fill those cracks. Gina just gets down on the ground with grief and wrestles with it, on her terms, until grief gives in and slinks off.

I worry about whether Harold ever found the bag that held his hands. Gina watches Nellie playing in our backyard. I worry about what lies beyond the grave. Gina watches Nellie playing in our backyard. I might have chosen to have no second dog rather than risk the pain of Nell's death ever again. Four months after Nellie died, Gina brought home two dogs and she still watches for Nellie in our backyard.

———

More than 99 percent of the cells that I called Harold may have survived his death. And 100 percent of the water that was 75 percent of Harold is definitely here. All of the elements that made up our friend Melanie and each of her proteins, each of her carbohydrates, each of her fats, each strand of her DNA—all of the molecules that were her—survive. Every breath our neighbor Phil ever drew is still here.

My uncle Harold was a carpenter. A good carpenter, I've been told. He had a special way with wood—a way with awls and augers, and gouges, and yes, saws. He learned cabinetry from his father. Harold spent a lot of time away from people, a lot of time with wood. Not long before his death, Harold made a pair of candlesticks for my mother—two wooden cylinders lathed to sensuous spirals thick as Harold's wrists. For as long as I can remember my mother kept her father's silver pocket watch atop one of those candlesticks and her mother's gold barrette on the other.

The wood is mahogany, and those candlesticks are all that is now left of a log my grandfather Isaac carried from his grandfather's house in Illinois clear to Kansas. Isaac had sawed the wood into three equal pieces. Two of those pieces were given to Harold, and the other to Harold's older brother James.

I've asked, but I can't find anyone who remembers why Isaac carried that log all that way or sawed it into three pieces. And no one I know remembers why he gave those three pieces to just two of his seven sons. Clearly, it was terribly important to Isaac or he would never have carried such a weighty piece of wood all those miles. No one knows what happened to the other piece of that log, the piece he gave to James. What I know, I know only because Harold once wrote on a slip of paper not much larger than his thumb that his father had given two pieces of that wood to him and one to James. The rest is gone. James is dead, Harold is dead, Isaac is dead. No one remembers.

———

The only thing we know for sure disappears when people die is something we can never know for certain was there to begin with—their personal knowledge. The things that only she remembered, the things that only he knew. Those are gone from us. But the impact of that loss on us, the living, is a little hard to assess since we never knew of any of these things anyway. What the dead alone knew is forgotten. Everything else remains.

Forgetting is another thing. Things forgotten are forgotten forever. Death leaves much to the imagination. The dead, all of them, or very nearly all of them, are still among us. I've seen them in coffee shops and on buses. Gina has seen them in our backyard.

Our immune systems, and only our immune systems, prevent us from becoming everyone else all at once. We are who we are only because we defend our selves every moment of every day. And who we are is everything.

We are pieces of others. Portraits painted somewhere between our brains and our thymuses. We are the dirt we've eaten and the songs we've sung. We are the light of stars and darknesses old beyond imagining. We are at once spontaneous fires and sacred water. We are faith and forgiveness. We are our own deaths and we are the eternal thoughts of others.

Colored dots on a broad blue canvas. Some born inside our bone marrow, some given to us by others. A few more that we gathered up on our own. Indestructible dots held by each of us for a lifetime. Dots we must give back at the end for those who cannot follow. Stand too close, and it looks like nothing more than scientific sleight of hand. Step back, and there is suddenly mitochondria, madness, faith, fatherhood, life, death, men, women.

Scientific discovery is as much about the discovery of new language or new uses for language as it is about anything else. New understanding demands new words, or at the very least that old words be twisted like warm iron to serve new purposes. Sometimes the words are hard. Sometimes they are not. But hard or soft, each of them paints another dot or two onto our canvases, enriches a little the colors of our lives and the complexity of the underlying mystery. We are all of those dots—the ones that science gave to us, the ones our parents gave to us, and the ones our uncles and lovers and children and dogs and cats gave to us—all of those dots all at once.

The pointillist person.

Inside of the dots are our words, inside of our words are our stories. Those stories are our greatest gifts—to ourselves, as well as to others. We are our stories.

SOURCES

"CHIMERA"

Andersson, G., A. Svensson, N. Setterblad, and L. Rask. "Retroviruses in the Human MHC Class II Region." *Trends in Genetics* 14:109 (1998).

Fauci, A., and D. L. Longo. "The Human Retroviruses." Chapter 192 in *Principles of Internal Medicine*. New York: McGraw Hill, 2000.

Janeway, C. A., P. Travers, M. Walport, and J. D. Capra. *Immunobiology*. New York: Garland Publishing, 1999. General immune function and immunological memory.

McHeyzer-Williams, M. G., and R. Ahmed. "B Cell Memory and the Long-lived Plasma Cell." *Current Opinion in Immunology* 11:172 (1999).

"Mysteries of the Mind." *Scientific American*, Special Issue, 1997.

Virgin, H. W., and S. H. Speck. "Unravelling Immunity to Herpesviruses: A New Model for Understanding the Role of Immunity in Chronic Virus Infections." *Current Opinion in Immunology* 11:371 (1999).

"SELF AND ANTISELF"

Broad, W., and N. Wade. "Deceit in History." Chapter 1 in *Betrayers of the Truth*. New York: Simon and Schuster, 1983.

Crick, F. *The Astonishing Hypothesis*. New York: Simon and Schuster, 1994.

Geison, G. L. *The Private Science of Louis Pasteur*. Princeton: Princeton University Press, 1995.

Janeway, C. A., P. Travers, M. Walport, and J. D. Capra. *Immunobiology*. New York: Garland Publishing, 1999.

Sapp, J. "The Nine Lives of Gregor Mendel." In *Experimental Inquiries*, edited by H. E. Le Grand. New York: Kluwer Academic Publishers, 1990.

"EATING DIRT"

dè Borhegyi, Stephen F. *El Santuario de Chimayo*. Santa Fe: Ancient City Press for the Spanish Colonial Arts Society, 1956.

Bottjer, S. W., and A. P. Arnold. "Developmental Plasticity in Neural Circuits for a Learned Behavior." *Annual Review of Neuroscience* 20:459 (1997).

Curtiss, S. *Genie: A Psycholinguistic Study of a Modern-Day "Wild Child."* New York: Academic Press, 1977.

Doupe, A. J., and P. K. Kuhl. "Birdsong and Human Speech: Common Themes and Mechanisms." *Annual Review of Neuroscience* 22:567 (1999).

Itard, J.-M.-G. *The Wild Boy of Aveyron*. Translated by G. Humphrey and M. Humphrey. New York: Appleton-Century-Crofts, 1962.

Lane, H. L. *The Wild Boy of Aveyron*. Cambridge, MA: Harvard University Press, 1976.

Lanning, D., P. Sethupathi, K. J. Rhee, S. K. Zhai, and K. L. Knight. "Intestinal Microflora and Diversification of the Rabbit Antibody Repertoire." *Journal of Immunology* 165:2012 (2000).

Paul, W., ed. "B Lymphocyte Development and Biology." *Fundamental Immunology*. Philadelphia: Lippincott-Raven, 1999.

Rook, G. A. W., and J. L. Stanford. "Give Us This Day Our Daily Germs." *Immunology Today* 19:113 (1998).

Rymer, R. *Genie*. New York: Harper Perennial, 1993.

"SELF-DEFENSE"

Bullock, B. A., and R. L. Henze. "Focus on Pathophysiology." In *Neurobiology of Psychotic, Mood, and Anxiety Disorders*. Philadelphia: Lippincott Williams & Wilkins, 2000.

Chambers, D. A., and K. Schauenstein. "Mindful Immunology: Neuroimmunomodulation." *Immunology Today* 21:168 (2000).

Downing, J. E. G., and J. A. Miyan. "Neural Immunoregulation: Emerging Roles for Nerves in Immune Homeostasis and Disease." *Immunology Today* 21:281 (2000).

Siegel, G. J., B. W. Agranoff, R. Wayne, Albers, S. K. Fisher, and M. D. Uhler. "Basic Neurochemistry: Molecular, Medical and Cellular Aspects." In Part Seven *Neural Processing and Behavior*. Philadelphia: Lippincott Williams & Wilkins, 1999.

Straub, R. H., J. Westermann, J. Schölmerich, and W. Falk. "Dialogue Between the CNS and the Immune System in Lymphoid Organs." *Immunology Today* 19:409 (1998).

Wilder, Ronald L. "Neuroendocrine-Immune System Interactions and Autoimmunity." *Annual Review of Immunology* 13:307 (1995).

"LIGHT AND SHADOW"

Dodelson, S., E. I. Gates, and M. S. Turner. "Cold Dark Matter." *Science* 274:69 (1996).

Moore, R. Y. "A Clock for the Ages." *Science* 284:2102 (1999).

Ostriker, J. P. "New Light on Dark Matter." <http://www.psc.edu/science/Ostriker/ostriker.html>

Pennisi, E. "Multiple Clocks Keep Time in Fruit Fly Tissue." *Science* 278:1560 (1998).

Rowan, L., and R. Stone. "Portrait of a Galaxy." *Science* 287:61 (2000).

"WATERMARKS"

Alberts, B., D. Bray, J. Lewis, M. Raff, K. Roberts, and J. D. Watson. *The Molecular Biology of the Cell.* New York: Garland Publishing, 1994.

Ball, P. *Life's Matrix: A Biography of Water.* Berkeley: University of California Press, 2001.

Blunier, T. " 'Frozen' Methane Escapes from the Sea Floor." *Science* 288:68 (2000).

Buchheim, J. "Oceanography: Water, Seawater, and Ocean Circulation and Dynamics." <http://www.marinebiology.org/oceanography.htm>

Haq, B. U. "Methane in the Deep Blue Sea." *Science* 285:543 (1999).

"Lourdes." Official Web site. <http://www.lourdes-france.com/bonjour.htm>

Nisini, B. "Water's Role in Making Stars." *Science* 290:1513 (2000).

"THE FLAME WITHIN"

Alberts, B., D. Bray, J. Lewis, M. Raff, K. Roberts, and J. D. Watson. *The Molecular Biology of the Cell.* New York: Garland Publishing, 1994.

"FBI Debunks Spontaneous Human Combustion." <http://www.apbnews.com/media/gfiles/humancombust/index/htm>

Gray, M. W., G. Burger, and B. Fritz Lang. "Mitochondrial Evolution." *Science* 283:1476 (1999).

Nelson, D. L., and M. M. Cox. *Lehninger Principles of Biochemistry.* 3rd ed. New York: Worth Publishers, 2000.

Tanford, C. *Physical Chemistry of Macromolecules.* New York: John Wiley & Sons, 1961.

"MADNESS"

Hatalski, C. G., A. J. Lewis, and W. I. Lipkin. "Borna Disease." *Emerging Infectious Disease* 3:129 (1997).

Swedo, S. E., et al. "Pediatric Autoimmune Neuropsychiatric Disorders Associate with Streptococcal Infections: The First 50 Cases." *American Journal of Psychiatry* 155:264 (1998).

Travis, J. "Undesirable Sex Partners: Bacteria Manipulate Reproduction of Insects and Other Species." *Science News* 150, no. 20 (1996). Also available at <http://www.sciencenews.org/sn_arch/11_16_96/bob1.htm>

Weschler, L. *Mr. Wilson's Cabinet of Wonder: Pronged Ants, Horned Humans, Mice on Toast and Other Marvels of Jurassic Technology.* New York: Vintage Books, 1996.

Zimmer, C. "Parasites Make Scaredy-Rats Foolhardy." *Science* 289:525 (2000).

"ACORNS OF FAITH"

Davies, Pete. *The Devil's Flu: The World's Deadliest Influenza Epidemic and the Scientific Hunt for the Virus That Caused It.* New York: Henry Holt & Co., 2000.

Dawkins, R. *The River Out of Eden: A Darwinian View of Life.* New York: Basic Books, 1995.

Hamilton, W. D. "Extraordinary Sex Ratios." *Science* 156:477 (1967).

Kierkegaard, Søren. *Fear and Trembling.* New York: Penguin Books, 1985.

Kolata, G. *Flu: The Story of the Great Influenza Pandemic of 1918 and the Search for the Virus That Caused It.* New York: Simon and Schuster Trade Paperbacks, 2001.

Lederberg, J. "Infectious History." *Science* 288:287 (2000).

"FORGIVING THE FATHER"

Alberts, B., D. Bray, J. Lewis, M. Raff, K. Roberts, and J. D. Watson. *The Molecular Biology of the Cell*. New York: Garland Publishing, 1994.

Bianchi, D. W. "Fetomaternal Cell Trafficking: A New Cause of Disease." *American Journal of Medical Genetics* 91:22 (2000).

Brown, R. E. "What Is the Role of the Immune System in Determining Individually Distinct Body Odours?" *International Journal of Immunopharmacology* 17:655 (1995).

Janeway, C. A., P. Travers, M. Walport, and J. D. Capra. *Immunobiology*. New York: Garland Publishing, 1999.

Helmuth, L. "Mom's Cells Tied to Autoimmune Ills." *Science News* 155 no. 18 (1999).

Nelson, J. L. "Autoimmune Disease and the Long-term Persistence of Fetal and Maternal Microchimerism." *Lupus* 8:493 (1999).

Piotrowski, P., and B. A. Croy. "Maternal Cells Are Widely Distributed in Murine Fetuses in Utero." *Biology of Reproduction* 54:1103 (1996).

Sophocles. *Oedipus Trilogy*. Project Gutenberg at <http://www.promo.net/pg/>

Wedekind, C., T. Seebeck, F. Paepke, and A. J. Paepke. "MHC-Dependent Mate Preferences in Humans." *Proceedings of the Royal Society of London, B, Biological Sciences* 260:245 (1995).

Wedekind, C., and S. Furi. "Body Odour Preferences in Men and Women: Do They Aim for Specific MHC Combinations or Simply Heterozygosity?" *Proceedings of the Royal Society of London, B, Biological Sciences* 264:1471 (1997).

"SAVED BY DEATH"

de Lange, T. "Telomeres and Senescence: Ending the Debate." *Science* 279:334 (1998).

Gould, S. J. *Full House: The Spread of Excellence from Plato to Darwin*. New York: Harmony Books, 1996.

Grant, M. C. "The Trembling Giant." *Discover* 14 no. 10 (1993).

Thompson, C. B. "Apoptosis." In *Fundamental Immunology*, edited by W. E. Paul. Philadelphia: Lippincott-Raven, 1999.

Woolf, N. *Pathology: Basic and Systemic*. London: W. B. Saunders, 1998.

"THE METAMORPHOSIS"

Ball, P. *Life's Matrix: A Biography of Water*. Berkeley: University of California Press, 2001.

Kingsley, R. E. *Concise Text of Neuroscience*. Philadelphia: Lippincott Williams & Wilkins, 1999.

Mims, C., J. Playfair, I. Roitt, D. Wakelin, and R. Williams. "The Organisms." Chapter 3 in *Medical Microbiology*. St. Louis: Mosby, 1998.

Web Elements. <http://www.webelements.com> Online periodic table with information about all of the elements.

Whitman, W. B., D. C. Coleman, and W. J. Wiebe. "Prokaryotes: The Unseen Majority." *Proceedings of the National Academy of Sciences* 95:6578 (1998).

DATE DUE